Software Engineering Construction Knowledge Areas

Volume 3

The Engineering of Software Projects

The Engineering of Software Projects
Software Engineering Knowledge Areas

These twelve volumes support the IEEE *Guide to Software Engineering Body of Knowledge (SWEBOK)* and the IEEE *Computer Society Professional Software Engineering Master Certification.*

This is a work-in-progress; each of these volumes is not currently available but should be published in the coming year.

—Richard Hall Thayer

Volume 1 — Software Engineering Requirements

Volume 2 — Software Engineering Design

Volume 3 — Software Engineering Construction

Volume 4 — Software Engineering Testing

Volume 5 — Software Engineering Maintenance

Volume 6 — Software Engineering Configuration Management

Volume 7 — Software Engineering Management

Volume 8 — Software Engineering Processes

Volume 9 — Software Engineering Models and Methods

Volume 10 — Software Engineering Quality Assurances

Volume 11 — Software Engineering Economics

Volume 12 — Software Engineering Project Management

These volumes will be published and sold through Amazon Books.

Software Engineering Construction Knowledge Areas

Volume 3
The Engineering of Software Projects

Richard Hall Thayer, PhD, CSDP

Contributing authors:

Steve McConnell, Construx Software
Mark J. Christensen, PhD, Independent Consultant

Software Management Training
Carmichael, California
2017

Table of Contents
Software Construction Knowledge Areas

A Partial List of General Abbreviations
(One-of-a-kind abbreviations are identified in place)

a.k.a.	—	also known as
API	—	application programming interface
a.s.a.p.	—	as soon as possible
ConOps	—	concept of operations (document)
CSCP	—	Computer Society Certificates of Proficiency
DFD	—	data flow diagram
FSM	—	finite state machines
HCI	—	human computer interface
HW	—	hardware
I/O	—	input/output
IDE	—	integrated development environment
IV&V	—	independent verification and validation
KA	—	knowledge area
PSEM	—	Professional Software Engineering Master (Certification)
SCM	—	software configuration management
SED	—	software engineering design
SEM	—	software engineering management
SEPM	—	software engineering project management
SER	—	software engineering requirements
SET	—	software engineering testing
SQA	—	software quality assurance
SW	—	software
SRE	—	software requirements engineering
SWE	—	software engineering
SWEBOK	—	(Guide to the) Software Engineering Body of Knowledge
TBD	—	to be determined/done
V&V	—	verification and validation

Honorary Foreword

To explain the origin of the term "software engineering,"
the following story is offered.[1]

In the mid-1960s, there was increasing concern in scientific quarters of the Western world that the tempestuous development of computer hardware was not matched by appropriate progress in software development. The software situation looked to be more turbulent. Operating systems had just become the latest rage, but they showed unexpected weaknesses. The uneasiness had been articulated in the NATO Science Committee by its U.S. representative, Dr. I.I. Rabi, the Nobel laureate and famous, as well as influential, physicist. In 1967, the Science Committee set up the Study Group on Computer Science, with members from several countries, to analyze the situation.

The German authorities nominated me for this team. The study group was given the task of "assessing the entire field of computer science," with particular elaboration on the Science Committee's consideration of organizing a conference and, perhaps, later, establishing an International Institute of Computer Science.

The study group, concentrating its deliberations on actions that would merit an international rather than a national effort, discussed all sorts of promising scientific projects. However, it was rather inconclusive on the relation of these themes to the critical observations mentioned above, which had guided the Science Committee in creating the study group.

Perhaps not all members of the study group had been properly informed about the rationale for its existence. In a sudden mood of anger, I remarked, "The whole trouble comes from the fact that there is so much tinkering with software. It is not made in a clean fabrication process." When I found out that this remark was shocking to some of my scientific colleagues, I elaborated on the idea with the provocative saying, "What we need is *software engineering.*"

This remark caused the expression "software engineering," which seemed to some to be a contradiction in terms, to be stuck in the minds of the members of the group. In the end, in late 1967, the study group recommended that we hold a Working Conference on Software Engineering, and I was made chairman. I not only had the task of organizing the meeting (which was held from October 7 to October 10, 1968, in Garmisch, Germany), but I had to set up a scientific program for a subject that was suddenly defined by my provocative remark.

---

1. Dr. Bauer originally wrote this paper as an introduction to a 1993 IEEE tutorial: *Software Engineering: A European Perspective*, R.H. Thayer, and A.D. McGettrick, eds., IEEE Computer Society Press, Los Alamitos, CA, 1993.

I enjoyed the help of my co-chairmen, L. Bolliet from France and H.J. Helms from Denmark. In addition, I had the invaluable practical support of the program committee members, A.J. Perlis and B. Randell in the section on design, P. Naur and J.N. Buxton in the section on production, and K. Samuelson, B. Galler, and D. Gries in the section on service.

Among the 50 or so participants, E.W. Dijkstra was dominant. Not only did he make cynical remarks like, "The dissemination of error-loaded software is frightening", and, "It is not clear that the people who manufacture software are to be blamed. I think manufacturers deserve better, more understanding users," but he also said, at this early date, "Whether the correctness of a piece of software can be guaranteed or not depends greatly on the structure of the thing made." He had very fittingly named his paper "Complexity Controlled by Hierarchical Ordering of Function and Variability," introducing a theme that followed his life over the next 20 years. Some of his words have become proverbs in computing, like "Testing is a very inefficient way of convincing oneself of the correctness of a program."

With the wide distribution of the reports from the Garmisch Conference and in a follow-up conference in Rome, from October 27 to 31, 1969, it happened that not only the term "software engineering" but also the idea behind this term became fashionable. Chairs were created, institutes were established (although the one that the NATO Science Committee had proposed did not come about because of reluctance on the part of Great Britain to have it organized on the European continent), and a great number of conferences were held.

The tutorial nature of the papers in this book is intended to offer readers an easy introduction to the topics and indeed to the attempts that have been made in recent years to provide them with the *tools,* both in a handcraft and an intellectual sense, which allow them now to honestly call themselves *software engineers.*

Friedrich L. Bauer, PhD
Professor Emeritus
Technische Universität München (TUM)
Germany

> P.S. *In 1989, I met Dr. Friedrich L. Bauer, Professor Emeritus, Universität München, when I was delivering a software engineering seminar in Munich for the IEEE. Later, Professor Bauer provided me with the story of how he came to name what we now call software engineering (I reprinted the story as an Honorary Foreword by Professor Bauer). I thought some of the readers, especially the younger ones, might be interested in the history of the naming of software engineering. Professor Bauer died last year (2015) at the age of 90. RHT*

Preface

This knowledge area defines and presents software construction practices and processes. It defines the essential processes and tools of software construction to identify the role of these processes and tools in the development of a functioning software system.

University students as well as candidates for the IEEE *Computer Society* Professional Software Engineering Master Certification exam in *software construction* need to focus on the following subareas of the design knowledge areas [https://www.computer.org/web/education/se-construc ti on].

The examination requires that a candidate demonstrate proficiency in:

1. *The managerial skills required in software construction complexity: anticipating change, constructing for verification, and employing construction standards throughout a construction project.*

2. *Applying key construction life cycle models—including planning and measurement—to software construction projects.*

3. *Managing construction considerations such as design, languages, coding, testing, quality, and reuse.*

4. *Using key construction technologies such as API, state-based and table-driven techniques, runtime configuration, middleware, and other technologies in a typical software construction project.*

5. *Working with key software construction tools such as GUI builders, unit testing tools, profiling, performance analysis, and slicing tools.*

Important Information for the Reader

Regarding references to SWEBOK in this book, when the second edition of the SWEBOK *Guide* is referenced, it is labeled [SWEBOK 2004]; accordingly, when the third edition of the SWEBOK *Guide* is referenced, it is labeled [SWEBOK 2014]. A review of both of these books shows that the technical material in these two volumes is many times, word for word, identical.

I need to point out that SWEBOK 2004 is, for all practical purposes, not copyrighted. In contrast, SWEBOK 2014 is copyrighted. Therefore, it is expedient for me to show my references, when identical in both SWEBOKs, as [SWEBOK 2004] as long as I comply with IEEE's usage limitations.

Another good reason for using SWEBOK 2014 (e.g., SWEBOK 2004) as a reference is its usefulness as an IEEE PSEM Certification exam study guide. Question writers tend to use the same wording as the question source. Since the question source is SWEBOK 2014, this will give the exam taker a "leg-up" in passing the IEEE PSEM Certification exam.

To accommodate both groups—university students as well as candidates for the IEEE Computer Society Certificate of Proficiency exam in *software requirements*—a software engineering principle that is not included in SWEBOK 2014 and is <u>not</u> likely to produce an exam question is marked with the following statement. "*Note: SWEBOK does not include <u>(to be filled in)</u> in the SWEBOK guide*. The certificate candidate is free to skip this entry.

However, this principle can be important to a university student-studying software engineering.

This volume uses the outline (paragraph headings) expressed in *Software Construction*, Chapter 3 *of Guide to Software Engineering Body of Knowledge (SWEBOK)*, IEEE Computer Society, 2014 as the outline for Volume 3.

Our book makes maximum use of SWEBOK—a very impressive document—and should be read by anybody studying software engineering.

This small book is divided into five parts.

(1) Part One presents an analysis of the appropriate software engineering knowledge areas (KAs) followed by an explanation of the material related to the topics contained in the SWEBOK.

(2) Two additional articles from highly regarded technical sources comprise Part Two:

    a. "Software Engineering Construction," by Steven McConnell, the well-known author of *Code Complete*.

    b. "Software Construction: Design, Write and Test the Code ," by Dr. Mark Christensen, previously published in *Software Engineering Essentials*, by Richard Hall Thayer and Merlin Dorfman, Amazon, 2012.

(3) The Third Part of this volume includes a discussion concerning *coding standards*—the need for coding standards and some possible solutions presented by Jordan Belone.

(4) Twenty sample exam questions that should help both the exam takers and the university students are included in Part Four.

(5) A Microsoft Word index.

Richard Hall Thayer, PhD, CSDP
Life Fellow of the IEEE
Member of the IEEE Computer Society Golden Core
Emeritus Professor of Software Engineering,
  Sacramento State University, California

Acknowledgments

No successful endeavor has ever been undertaken by one person alone. I would like to thank the following people and organizations who supported me in this effort.

I first want to thank my wife Mildred for her high degree of tolerance in putting up with my working seven days a week on this manuscript. Without her encouragement and support, this book could never have been completed.

I also want to thank Ellen Sander who performed copyediting, Jon Digerness of North Coast Graphics for providing me with the comic illustrations, and Jim Tozza for giving me hardware and software support to keep my computer equipment running.

In addition, I want to thank Steve Tockey for providing me with tips about the Computer Society exam specifications in order to maximize the usefulness of our software engineering textbook and SWE exam guidebook, and Melville (Mel) Piercey of Copy Plus for providing cover artwork and designing and drawing the engineering chapter graphics.

Finally, I want to thank our little dog Maxwell (a.k.a. Max, Maxcito, or Speedy) who kept me company in the evening hours when everybody else had gone to bed.

A happy Max says that:

This is a Terrrrrrrific Book. I chewed on a copy,
and it was tasty.

A Note to Our Readers

One of the advantages of using a "print-on-demand" (POD) publishing service is the ability to make changes to the manuscript relatively easily when errors or improvements are identified.

The authors encourage you to identify and send potential errors or suggested improvements to the e-mail address listed below. I do not guarantee to make all the changes identified, but I do promise to review and seriously consider all recommendations.

Disclaimer

While I have more than 50 years of software engineering experience, including university teaching, I am not a technical expert in every component of software engineering. To make up for this shortcoming, I have made extensive use of material written by subject matter experts and papers (many posted on the web) as source documents.

Every effort has been made to make this software engineering reference as complete and accurate as possible. However, I can make no representation or warranties with respect to accuracy or completeness of the contents of this book and specifically disclaim any implied warrantee of merchantability or fitness for a particular purpose. The advice and strategies contained herein may not be suitable for your situation. If in doubt, you should consult with a professional software engineer. Where appropriate, neither I nor the printer will be liable for the loss of profit or other commercial damages including but not limited to special, incidental, consequential, or other damages [IEEE Press Disclaimer].

Please keep me posted.

Richard Hall Thayer, PhD, CSDP
thayer@csus.edu

Chapter 1
Software Engineering Construction Fundamentals

This chapter is a textbook and study guide introducing the principles and some of the problems of software engineering construction (SEC). This book can be used in a university software engineering course and as a study guide to aid individual software engineers in passing the IEEE Professional Software Engineering Master (PSEM) Certification exams in software construction.

This chapter introduces concepts and problems of SEC (some software engineers might call this "computer programming"). The term "software engineering construction" refers to the detailed creation of working and meaningful software through a combination of coding, verification, unit testing, integration testing, and debugging.

Software construction is linked to all the other software engineering efforts, most strongly to software design and software testing. Although some detailed design work may be completed prior to construction, much design is performed during construction, and the SEC process itself involves significant software design and test activity.

INTRODUCTION

The *software engineering construction knowledge area (KA)* refers to the detailed creation of meaningful software through a combination of coding, verification, unit, and integration testing, and debugging. This KA includes five knowledge subareas:

(1) ***Software Construction Fundamentals*** introduces the basic principles of construction: minimizing complexity, anticipating change, constructing for verification, and standards for construction.

(2) ***Managing Construction*** describes the topics of construction models, construction planning, and construction measurement.

(3) ***Practical Considerations*** presents construction design, construction languages, coding, construction testing, reuse, construction and code quality, and integration.

(4) ***Construction Technologies*** includes run-time issues, documentation, error handling, performance analysis, timing, and test-first programming.

(5) ***Software Construction Tools*** introduces some of the tools used in constructing software.

"Construction" is the building of something, usually involving managing, designing, building, and checking the final product. *Software construction* is a software engineering discipline. It is the detailed creation of working meaningful software through a combination of coding, verification, unit testing, integration testing, and debugging.

Software construction is linked to all the other software engineering efforts, most strongly to software design and software testing. This linkage exists because the SEC process itself involves significant software design and test activities. It also uses the output of design and provides one of the inputs to testing. Detailed boundaries between design, construction, and testing will vary depending upon the software life cycle processes and methods that are used in a project [SWEBOK 2004].

This chapter is important to PSEM Certification exam takers and to university software engineering students. The major reference (and one might say the only reference) is the second edition of Steve McConnell's book, *Code Complete* [2004]. This excellent book is very easy to read and understand. I will be using Mr. McConnell's ideas in this chapter, (with Mr. McConnell's permission). Mr. McConnell uses the term "construction" because he believes the terms "coding" and "implementation" imply mechanical translations from design to machine language. "Construction" provides for creativity and judgment.

Construction entails mostly coding and debugging, but it also involves limited SW requirements, detailed design, construction planning, unit testing, integration and integration testing, and other activities.

BREAKDOWN OF TOPICS FOR SOFTWARE CONSTRUCTION

Software construction is a software engineering discipline. It is the detailed creation of working and meaningful software through a combination of coding, verification, unit testing, integration testing, and debugging. It is linked to all the other software engineering disciplines, most importantly to software design and testing. Figure 1 provides a top-level decomposition and breakdown of the SEC KA. Figure 2 illustrates a sequence of software development phases and their relationships with each other; the phases and products associated with SEC are marked with a "star."

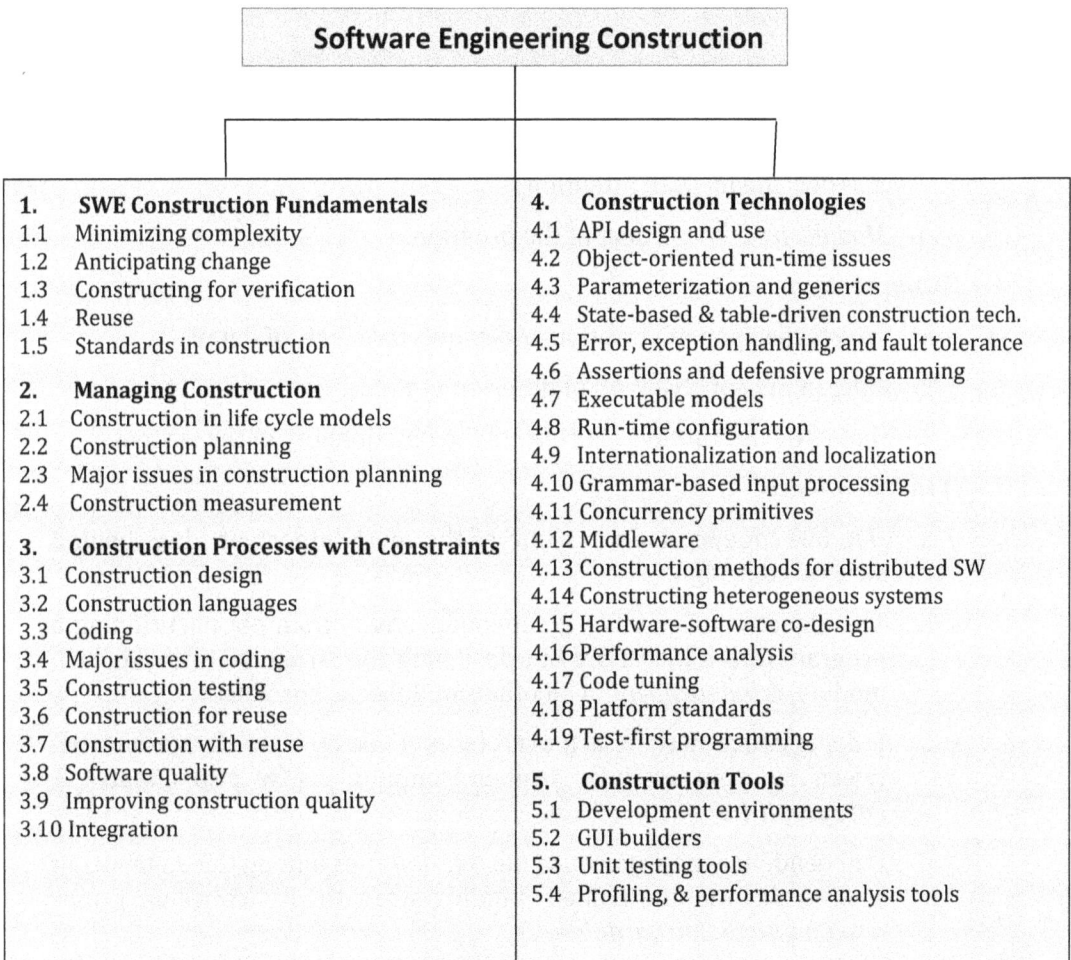

Software Engineering Construction

1. SWE Construction Fundamentals
1.1 Minimizing complexity
1.2 Anticipating change
1.3 Constructing for verification
1.4 Reuse
1.5 Standards in construction

2. Managing Construction
2.1 Construction in life cycle models
2.2 Construction planning
2.3 Major issues in construction planning
2.4 Construction measurement

3. Construction Processes with Constraints
3.1 Construction design
3.2 Construction languages
3.3 Coding
3.4 Major issues in coding
3.5 Construction testing
3.6 Construction for reuse
3.7 Construction with reuse
3.8 Software quality
3.9 Improving construction quality
3.10 Integration

4. Construction Technologies
4.1 API design and use
4.2 Object-oriented run-time issues
4.3 Parameterization and generics
4.4 State-based & table-driven construction tech.
4.5 Error, exception handling, and fault tolerance
4.6 Assertions and defensive programming
4.7 Executable models
4.8 Run-time configuration
4.9 Internationalization and localization
4.10 Grammar-based input processing
4.11 Concurrency primitives
4.12 Middleware
4.13 Construction methods for distributed SW
4.14 Constructing heterogeneous systems
4.15 Hardware-software co-design
4.16 Performance analysis
4.17 Code tuning
4.18 Platform standards
4.19 Test-first programming

5. Construction Tools
5.1 Development environments
5.2 GUI builders
5.3 Unit testing tools
5.4 Profiling, & performance analysis tools

Figure 1: Hierarchical listing of topics for SWE construction KAs

1. Software Engineering Construction Fundamentals

1.1 Minimizing complexity.

A major factor in SEC is the inability of people (i.e., developers) to hold complex structures and information in their working memories, especially over long times. This leads to one of the strongest issues of SEC, that of *minimizing design complexity*. The need to reduce complexity applies to essentially every aspect of SEC and is particularly critical to the process of verification and testing of SECs. A number of approaches are listed below:

(1) Partitioning:

    a. *Reduces* the number of factors that must be dealt with simultaneously.

    b. *Splits* the system into components; this may increase complexity if not applied properly.

    c. *Produces* well-defined, documented boundaries in a program that narrows the focus of attention.

    d. *Modularizes* the product of the program.

(2) Hierarchy:

    a. *Directs* the span of attention and allows for levels of detail.

    b. *Aides* in the construction of different systems.

    c. *Shapes* the hierarchy that reflects the functionality of the system.

(3) Independence:

    a. "Module independence" is one of the most important ideas behind structured design.

    b. The objective of module independence is not simply partitioning a program into a hierarchy but designing the structure to make each module as *independent* of all other modules as possible.

    c. Independence may be achieved by *minimizing the relationships* between different modules (called "coupling"). (*See also Chapter 3, Paragraph 4.4)*

    d. Independence may *maximize the relationships* among the elements of a particular module (called "binding strength" or "cohesion"). (*See also Chapter 3, Paragraph 4.4.*)

    e. Independence masks each module view and knowledge of the remainder of the system as much as possible. This is the principle of "information hiding" [Parnas 1972], or a "need to know" basis of communication. (*See also Chapter 3, Paragraph 6.2.*)

1.2 Anticipating change.

Most software will change over time and the anticipation of change drives many aspects of SEC. Software is part of the changing external environments and changes in those outside environments affect software in many diverse ways; the need to *anticipate change* is supported by many specific construction techniques, in particular by communication methods, programming languages, platforms, and tools.

(1) ***Communication methods*** — For example, standards for document formats and contents.

(2) ***Programming languages*** — For example, language standards for programming languages such as Java and C++.

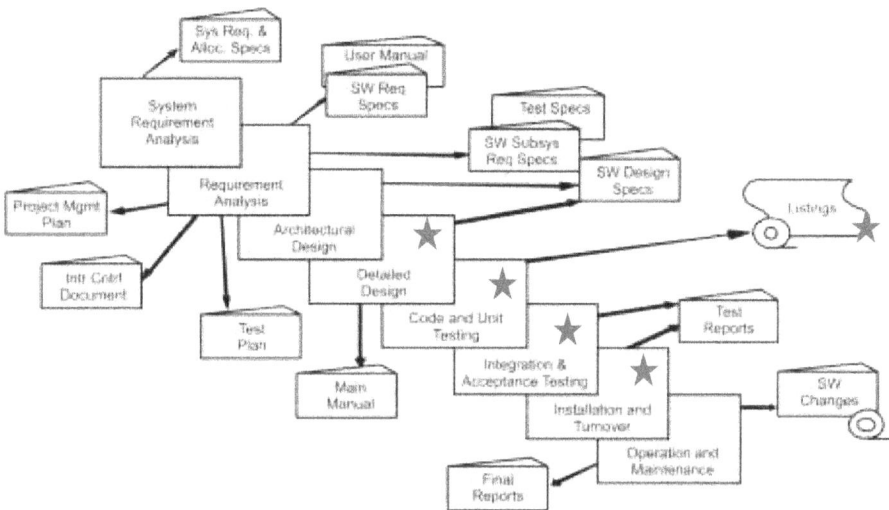

Figure 2: Sequence of software development phases

(3) ***Platforms*** — For example, programmers interface standards for operating system calls.

(4) ***Tools*** — For example, diagrammatic standards for notations like the Unified Modeling Language (UML).

1.3 Constructing for verification.

Constructing for verification means building software in such a way that faults can be readily diagnosed by the software engineers while writing the software, and during the later stages of independent testing and operational activities. Specific techniques that support constructing for verification include following coding standards to support code reviews and unit testing, organizing code to support automated testing, and restricting the use of complex or hard-to-understand language structures, and so forth [SWEBOK 2004].

1.4 Reuse.

SWE reuse is the idea that computer products written for one purpose can be, should be, or are being used in another program written later. (*See also Paragraphs 3.6 and 3.7.*) These existing assets can be used when solving different problems (i.e., can be *reused*). In SEC, typical assets that are reused include libraries, modules, components, source code, and commercial off-the-shelf (COTS) assets. Reuse is most effective when practiced systematically according to a well-defined, repeatable process. Systematic reuse can enable significant software productivity, as well as quality and cost improvements [SWEBOK 2014].

Reuse has two closely related facets: "construction *for* reuse" and "construction *with* reuse." The former means to create reusable software assets, while the latter means to reuse software assets in new construction.

Ad hoc code reuse has been practiced from the earliest days of programming. Programmers have always reused sections of code, templates, functions, and procedures that they personally developed. Software reuse as a recognized area of study in software engineering, however, dates only from 1968, when Douglas McIlroy of Bell Laboratories proposed basing the software industry on reusable components.

The *software library* is a primary example of code reuse. Programmers may decide to create internal abstractions so that parts of their program can (1) be reused or (2) create custom libraries for their own use. Some characteristics that make software more easily reusable are *modularity, loose coupling, high cohesion, information hiding,* and *separation of concerns.* (For definitions of these software engineering concepts *see Chapter 1, Paragraph 4.2; Chapter 3, Paragraphs 4.4 and 6.2)*

For newly written code to use a piece of existing code, some kind of interface, or means of communication, must be defined. These commonly include a "subroutine call" or use of a subroutine, object, class, or prototype. In organizations, such practices are formalized and standardized by domain engineering, a.k.a. a *software product line.*

Types of reuse can include [http://en.wikipedia.org/wiki/Code_reuse]:

> (1) ***Opportunistic*** — While getting ready to begin a project, the team (typically team members who developed the original components) realizes that existing components can be reused.
>
> (2) ***Planned*** — A team strategically designs components to be reusable in future projects or even in the same projects.
>
> (3) ***Internal reuse*** — A team reuses its own components. This may be a

business decision made when a team wants to control a component critical to the project.

(4) *External reuse* — A team may choose to license a third-party component. Licensing a third-party component typically only costs the team a small percentage of the cost to develop the component internally. However, the team must also consider the time required to find, learn, and integrate the component into the existing system

1.5 Standards in construction.

Construction standards refer to the application of *external* or *internal development* quality and cost. Specifically, the choices of allowable programming language subsets and usage standards are important aids in achieving higher software security. Standards that directly affect construction issues include the following [SWEBOK 2004]:

(1) *Use of external standards* — Construction depends on the use of *external standards* for construction languages, construction tools, technical interfaces, and interactions between SEC and other KAs. Standards come from numerous sources, including hardware and software interface specifications such as an Object Management Group (OMG) and international organizations such as the IEEE, the ISO, and the IEC.

The *International Organization for Standardization (ISO)* (English title), or L'*Organisation Internationale de Normalisation* (French title), is an international standard-setting body composed of various national standards organizations (including the IEEE). The ISO's primary products are published in the form of international standards. The ISO also publishes tech reports, technical specifications and other references [http://en.wikipedia.org/wiki/International_Organization_or_Standardization].

A second standard organization for software engineers is the International Electrotechnical Commission (IEC) (English title), or *La Commission Électrotechnique Internationale (CEI)* (French title). The IEC is a nonprofit group that publishes international standards for all electrical, electronic, and related technologies—collectively known as "electrotechnology."

These organizations were engaged primarily because no one know in which discipline software engineering standards belonged. In other words, it was neither "fish nor fowl," (i.e., something that is not easily categorized). Software engineering standards are now under the control of a joint ISO/IEC committee. This committee developed ISO/IEC 12207:2008, *Systems and Software Engineering—Software Life Cycle Processes,* an international standard for software life cycle processes. It aims to be the standard that defines each task required for developing and maintaining a software system. These two standards organizations are now *jointly* responsible for software engineering standards.

The IEEE is one of the early developers of software engineering standards. Fletcher Buckley, the original chair of the IEEE Computer Society Technical Committee for Software Engineering, initiated the software engineering standards effort.

Richard (Dick) Fairley and I led the team that developed IEEE Standard 1058-1998, *IEEE Standard for Software Project Management Plans.* In addition, with the help of Per Bjorke, a Norwegian graduate student at Sacramento State University, we developed IEEE Standard 1362-1998, *IEEE Guide for Information Technology – System Definition – Concept of Operations (ConOps).* (This standard was later used as a basis for a DoD standard.)

(2) ***Use of internal standards*** — Standards may also be created on an organizational basis, in particular at the corporate level for use on specific projects. These standards support coordination of group activities, minimizing complexity, anticipating change, and constructing for verification.

2. Managing Construction

The fundamentals of managing construction include *construction in life cycle models, construction planning*, and *construction measurement* [SWEBOK 2004].

2.1 Construction in life cycle models.

Numerous *construction models* have been created to develop software. Some models are linear from the construction point of view, such *as waterfall* and *staged-delivery* life cycle models. These models treat construction as an activity that occurs only after significant prerequisite work has been completed—including *detailed requirements, extensive design,* and *detailed planning.*

The linear approaches tend to emphasize the activities that precede construction (requirements design) and tend to create distinct separations between the activities. In these models, the main emphases of construction may be placed on coding [SWEBOK 2004].

(1) ***Waterfall life cycle model*** — The waterfall model (also called a "waterfall chart") is based on *The Waterfall Chart* by Winston W. Royce [1971] (*see Figure 3*). Royce's model establishes milestones, documents, and reviews at the end of each phase. This model:

   a. Establishes a baseline at periodic intervals.

   b. Uses configuration management to control the baselines.

   c. Performs iteration between phases.

d. Dr. Royce portrayed "his" waterfall chart as both a forward-looking flow of data (as each activity refines the project data) and a reverse flow of project data (to rework an earlier effort). Royce portrayed a "reverse" flow on his waterfall chart, called "iteration", to account for the return from a SWE activity to a previous activity, to correct an error, or to make a much-needed improvement in the overall system (see Figure 3).

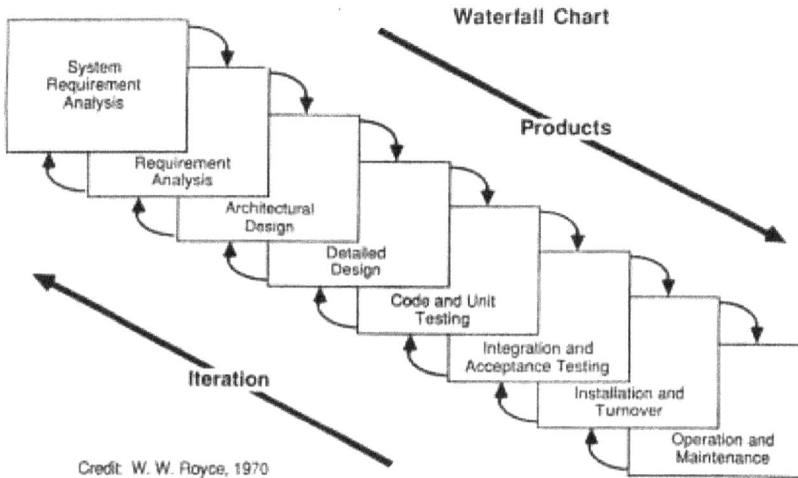

Figure 3: Waterfall chart

This software engineering process model is a popular technique used within the U.S. aerospace industry. This model is also called a *baseline management system, conventional process model, linear sequence model,* and *grand design development strategy* (draft Mil-Std. 498), and a *once-through development strategy* (final Mil-Std. 498).

Unfortunately, the U.S. government's version of the waterfall chart did not permit iteration between phases.

(2) ***Baseline management model*** — The *baseline model* is based on the work product (called a baseline) that has been formally reviewed and accepted as representing the product as developed at a particular point in time and serves as the basis for further development.

Each baseline must specify the:

a. Items that form the baseline (for example, *software requirements specifications*).

b. Reviews and approval mechanisms.

c. Acceptance criteria associated with the baseline.

d. Project organizations that participated in establishing the baseline.

Baselines are typically established at the completion of a major software engineering documentation effort (e.g., requirements specifications, detailed design description, and completed testing). Once established, a baseline will be changed for only one of two reasons: (1) the requirements have changed or (2) to correct a defect.

(3) **Staged-delivery models** — A *staged-delivery model* allows products to become deliverable at every stage. It prioritizes requirements set by project managers and these are clearly addressed as the developer proceeds along each stage. Additionally, creating a budget is reported to be easier using the staged-delivery model, in contrast to a software engineering document-driven system.

Other models are more iterative, such as *evolutionary prototyping, extreme programming,* and *Agile development.* These approaches tend either to treat construction as an activity that occurs concurrently with other software development activities, including requirements, design and planning, or to overlap them. These approaches tend to mix design, coding, and testing activities, and they often treat the combination of activities as construction [SWEBOK 2004]. Consequently, what is considered "construction" depends to some degree on the life cycle model used. Examples are:

(4) **Evolutionary prototyping** — An evolutionary prototype process model can be used when:

a. Requirements cannot be specified a priori.

b. The acquirer/developer/user wants to reserve some requirements decisions to be delivered later in the development cycle.

To use the evolutionary prototype model, the development engineer needs to establish their most accurate (but probably incomplete) estimate of the software requirements. The developer, along with the help of the potential user, will experiment with the partial system to develop a new set of requirements specifications that are more complete.

This cycle will be repeated until the developer and/or the user is satisfied that the software requirements are correct and complete (see Figure 4). When the system is completed, an SRS is written. Note that evolutionary prototyping must not be used as an excuse to "hack" a system without a SW requirements specification. Each cycle is a "mini-waterfall model" that uses the knowledge gained from the present cycle to generate requirements for the next cycle.

Cycle 1:

| Analyze | => | Design | => | Implement | => | Test | => | Evaluate |

Cycle 2:

| Analyze | => | Design | => | Implement | => | Test | => | Evaluate |

Cycle 3:

| Analyze | => | Design | => | Implement | => | Test | => | Evaluate |

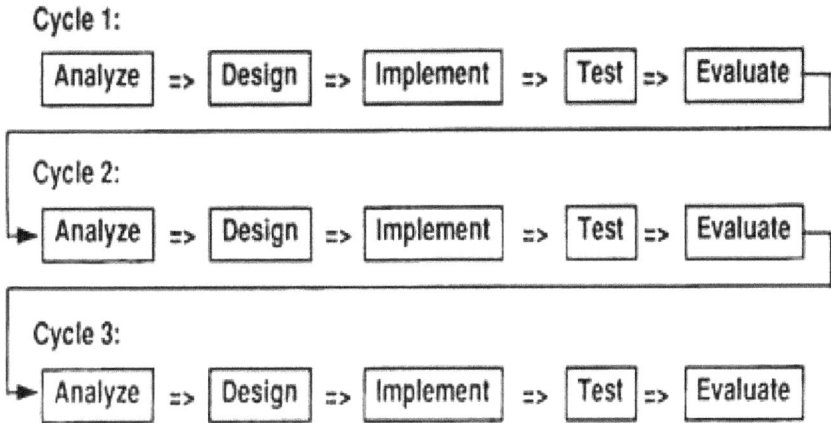

Figure 4: Evolutionary prototype model

The duration of each cycle must be limited. There are two schools of thought (supported by well-known software engineers) concerning the length of each cycle. The older school suggests a length of one to three months would be appropriate and the newer school suggests a length of two to four weeks. In either case, each cycle uses the knowledge gained from the present cycle to generate requirements for the next cycle. The next cycle might, or might not, use the software generated in the previous cycle.

This process ends when one of the following three endings is reached:

a. The users' needs are satisfied.

b. There can be a transition to an incremental approach.

c. The project is infeasible or cannot be finished.

(5) ***Agile development*** — Agile software development is a group of software development methods based on *iterative* and *incremental* development methods in which requirements and solutions evolve through collaboration between self-organizing, cross-functional teams. Agile promotes adaptive planning, evolutionary development, and timely delivery using a "time-boxed" iterative approach, and it encourages rapid and flexible response to change.

The Twelve Principles of Agile Processes are [http://scrummethodology .com/the-agile-manifesto-and-twelve-principles]:

1. Our highest priority is to satisfy the customer through early and continuous delivery of valuable software.

2. Welcome changing requirements, even late in development. Agile processes harness change for the customer's competitive advantage.

3. Deliver working software frequently, from a couple of weeks to a couple of months, with a preference to the shorter timescale.

4. Business people and developers must work together daily throughout the project.

5. Build projects around motivated individuals. Give them the environment and support they need and trust them to get the job done.

6. The most efficient and effective method of conveying information to and within a development team is face-to-face conversation.

7. Working software is the primary measure of progress.

8. Agile processes promote sustainable development. The sponsors, developers, and users should be able to maintain a constant pace indefinitely.

9. Continuous attention to technical excellence and good design enhances agility.

10. Simplicity—the art of maximizing the amount of work not done—is essential.

11. The best architectures, requirements, and designs emerge from self-organizing teams.

12. At regular intervals, the team reflects upon how to become more effective, then tunes and adjusts its behavior accordingly.

The *Agile Manifesto* was introduced in 2001. Since then, the *Agile Movement*, with its values, principles, methods, practices, tools, champions, and practitioners, philosophies, and cultures reported to have significantly changed the landscape of modern software engineering and commercial software development in the Internet era [http://en.wikipedia.org/wiki/Agile_software_development].

2.2 Construction planning.

The choice of construction method affects the extent to which construction prerequisites are performed, the order in which they are performed, and the completion date before construction work begins [SWEBOK 2004].

This approach to construction planning affects the project's ability to reduce complexity, anticipate change, and construct for verification. These objectives may be addressed at the process, requirements, and design levels, but they will also be influenced by the construction method [SWEBOK 2004].

Construction planning involves laying out the work plan (schedule) to design, implement, debug, and unit test the software. It also defines the order in which components are created and integrated, the software quality management processes, the allocation of task assignments to specific software engineers, and other tasks according to the chosen approach [SWEBOK 2004].

2.3 Major issues in construction planning.

(1) Coders are typically not good planners.

(2) Time will be very difficult to manage unless a well-developed architectural design is in place and used to estimate project SW size.

(3) Many organizations do not collect project data on which to base future estimates.

(4) Many companies and projects do not use systematic cost estimating methods (models).

(5) Many managers and software developers consider planning to be a waste of time, and therefore, discourage it. They believe that plans are just going to change, so "why waste our time?"

(6) Project plans may be limited to the coding phase.

2.4 Construction measurement.

Numerous construction activities and artifacts can be measured, including *code developed, code modified, code reused, code destroyed, code complexity, code inspection statistics, fault-fix and fault-find rates, effort,* and *scheduling.* These measurements can be useful for purposes of managing construction, ensuring quality during construction, and improving the construction process, as well as for other reasons [SWEBOK 2004]. This "bookkeeping" effort can be easily done by a project management assistant.

3. Construction Processes with Real-World Constraints

Construction is an activity in which the software must come to terms with sometimes arbitrary and chaotic real-world constraints. Due to its proximity to real-world constraints, construction is driven more by practical considerations than some other KAs. Software engineering is perhaps most craft-like (i.e., artistic) in the construction area.

Some practical processes in construction include *construction design, construction languages, coding, construction testing, construction for reuse, and construction with reuse, construction quality, documentation,* and *system integration* [SWEBOK 2004].

3.1 Construction design.

Some projects allocate more design activity to construction while others allocate activity to a phase explicitly focused on design.

Regardless of the exact allocation, some detailed design work will occur at the construction level, and that design work tends to be dictated by immovable constraints imposed by real-world problems that are addressed by the software being built. Just as construction workers building a physical structure must make small-scale modifications to account for unanticipated gaps in the builder's plans, SEC workers must make design modifications on a smaller or larger scale to flesh out details of software design during construction [SWEBOK 2004]. *Construction design* includes the following [McConnell 2004]:

(1) **Software design** — Defining the software architecture (structure), components, modules, interfaces, test approach, and data for a software system to satisfy specified requirements [IEEE Std. 610.12-1990]:

    a. **Architectural design** — Identifying the major subsystems and modules.

    b. **Detailed design** — Identifying all modules in the system, the services offered by each module, and allocating the design module into the system architecture.

(2) **Code design** — Laying out the detailed functionality of the individual routines. A typical example might use a program design language (PDL) to "rough out" the code.

(3) **Data design** — Includes the following [McConnell 2004]:

    a. Make full use of data types.

    b. Create your own individual data types when allowed by the language.

    c. Create types with functionally oriented names.

    d. Choose explicated variable names.

    e. Use variables properly.

    f. Understand the scope of the variable.

    g. Use global variables properly.

    h. Use fundamental data types properly.

    i. Avoid the use of pointers.

3.2 Construction languages.

Construction languages include all forms of communication by which a human can specify an executable problem solution to a computer [SWEBOK 2004]. The simplest type of construction language is a configuration language, by which software engineers choose from a limited set of predefined options to create new or custom software installations. The text-based configuration files used in both the Windows and UNIX operating systems is an example of a software installation, and the menu-style selection lists of some program generators constitute another type of software installation [SWEBOK 2004].

(1) **Toolkit languages** — These languages are used to build applications out of toolkits (integrated sets of application-specific reusable parts) and are more complex than configuration languages. Toolkit languages may be explicitly defined as application programming languages (for example, scripts), or they may be implied by the set of interfaces of a toolkit.

(2) **Programming languages** — Programming languages are the most flexible type of construction language. They also contain the least amount of information about application areas and development processes, and so require the most training and skill to be effectively used.

There are three general kinds of notations used for programming languages, namely *linguistic, formal,* and *visual* [SWEBOK 2004]:

a. **Linguistic notations** — These notations are distinguished in particular by the use of word-like strings of text to represent complex SECs, and the combination of such word-like strings into patterns that have a sentence-like syntax. Properly used, each such string should have a strong semantic connotation, providing an immediate intuitive understanding of what will happen when the underlying SEC is executed [SWEBOK 2004].

b. **Formal notations** — Formal notations rely less on intuitive, everyday meanings of words and text strings and more on definitions backed up by precise, unambiguous, and formal (or mathematical) definitions. Formal construction notations and formal methods are at the heart of most forms of system programming where accuracy, time behavior, and testability are more important than ease of mapping into natural language. Formal constructions also use precisely defined ways of combining symbols that avoid the ambiguity of many natural language constructions [SWEBOK 2004].

c. **Visual notations** — Visual notations rely much less on the text-oriented notations of both linguistic and formal construction, but instead rely on direct visual interpretation and placement of visual entities that represent the underlying software. Visual construction tends to be somewhat limited by the difficulty of making "complex" statements using only movement of visual entities on a display. However, it can be a powerful tool where the primary programming task is simply to build and "adjust" a visual interface for a program [SWEBOK 2004].

3.3 Coding.

The following considerations apply to the use of SEC coding activity [SWEBOK 2004]:

(1) Techniques for creating understandable source code, including naming and source code layout.

(2) Approaches for using classes, enumerated types, variables, named constants, and other similar entities.

(3) Methods of using control structures.

(4) Ways to handle error conditions—both planned errors and other types of errors—(e.g., input of bad data).

(5) Ways of preventing code-level security breaches (for example, buffer overruns or array index overflows).

(6) Ways of using resources via exclusion mechanisms and discipline in accessing serially reusable resources (including threads or database locks).

(7) Methods of organizing code into statements, routines, classes, packages, or other structures.

(8) Ways of using code documentation (*see Paragraph 3.8*).

(9) Methods of code tuning (*see Paragraph 4.17*).

McConnell's book encourages effective coding [McConnell 2004].

(10) McConnell suggests assigning two developers to every part of the project. Assigning developers to every area of design has many advantages.

    a. First, one can be assured that at least two developers think the code is adequate.

    b. This tactic presents a risk reduction that at least two individuals are familiar with each bit of the code.

(11) Senior developers should always approve the finished documents.

(12) The rewards structure should be aimed toward developers who consistently produce a working product, not just one-time "wonders" that arrive in the middle of the night to fix a problem that they probably created in the first place.

The design structure should also ensure the application of effective coding standards. Since the IEEE Computer Society has not published coding standards, they will have to be developed internally within each organization. Classically, coding standards are different for each programming language.

3.4 Major issues in coding.

The following eight issues are present when coding a software project.

(1) Coding is sometimes viewed (by managers and customers) as the only important and necessary activity when constructing a software system.

(2) Coding alone cannot scale-up to larger systems.

(3) Coding alone is considered less expensive than software engineering.

(4) Coding alone is difficult to modify and update.

(5) Coding alone is difficult to maintain.

(6) Coding is easy to learn (compared to software engineering).

(7) Coding may not include *unit testing.*

(8) Programmers do not like to be called "coders."

3.5 Construction testing.

Construction involves two forms of testing, which are often performed by the software engineer who wrote the code [SWEBOK 2004]. These are:

(1) **Unit testing** — *Unit testing* is the lowest level of testing. Unit testing is not a formal test *per se*, but a test conducted by the programmer/coder who coded the module. Unit testing is frequently not included in the software test plan and essentially implements *white box testing*

(2) **Integration testing** — Integration testing serves two purposes: (1) to define defects in the software design and (2) define defects in the interfaces between components. The tests are derived from the architectural design of the system.

The purpose of *construction testing* is to reduce the time during which faults are inserted into the code and the time those faults are detected. The construction tests are traditionally developed after the code itself is written, but some practitioners and some methods advocate writing the tests before coding is complete.

Construction testing typically involves a subset of testing types. For instance, construction testing does not typically include system testing, alpha testing, beta testing, stress testing, configuration testing, usability testing, or other more specialized kinds of testing.

3.6 Construction for reuse.

Construction for reuse can be a very useful approach. Generally, reuse is practiced by companies that plan to build a series of software systems around a common core (also called a "product line"). This approach is typically most effective when the series is carefully planned out; it would be very difficult to plan the series as an afterthought.

Software product lines were first initiated by manufacturers who had long employed analogous engineering techniques to create a product line consisting of similar products using a common factory that assembled and configured parts, which could be used across the product line. For example, automotive manufacturers can create unique variations of one car model using a single pool of carefully designed parts and a factory specifically designed to configure and assemble those parts.

The characteristic that distinguishes software product lines from previous efforts is *predictive* versus *opportunistic* software reuse. In predictive reuse, rather than putting general software components into a library in the hope that opportunities for reuse will arise, software product lines call for software artifacts to be created only when reuse is predicted in one or more products in a well-defined product line.

Recent advances in the software product-line field have demonstrated that narrow and strategic applications of these concepts can yield order of magnitude improvements in software engineering capability. The result is often a "spike" in competitive business advantages, similar to that seen when manufacturers adopt mass production and mass customization paradigms [http://en.wikipedia .org/wiki/Code_reuse].

3.7 Construction with reuse.

Construction with reuse is defined as the construction of a new software system with the use of existing software components. The new software system can use anywhere from a small component of an existing software system to the complete software system. The most difficult part of reusing software is identifying where to obtain the needed software and how to use it. Reuse is not just the use of a completed (and working) software system; it can involve any one of the following components [http://en.wikipedia.org/wiki/Code_reuse].

Some of these reusable components can be (listed according to ease of usability):

    (1) Project management plans.

    (2) Cost and schedule estimates. (This is essential to the practice of software cost and schedule estimating.)

    (3) Software documentation such as requirements specifications, quality assurance processes, and design descriptions.

    (4) Completed code (macros, statements, sections, programs, subsystems, etc.).

An original system type of reuse occurs when a software development organization undertakes a software development project (i.e., contract) to build a software system similar to one it had built earlier. (This generally works well only if most of the original developers are still available.) The general practice of using a prior version of an extant program as a starting point for the next version is also a form of program reuse.

A number of years ago, I witnessed an ill-advised attempt to generalize large-scale "reuse" in a very large, complex government organization. This organization attempted to institute a manual database of existing software programs (numbering in the tens-of-thousands) that would be available to internal organizations needing a software system for a given application.

The requesting organization would send a description of its software requirements to the parent organization office that was responsible for maintaining the catalog of existing systems. The data-list manager (only one person was assigned to the job) would then match the description of the new requirements to the available software systems and send the requesting organization a copy of the matched system. This matching had to be attempted before the parent organization would approve the development of a new system.

To my knowledge, there was never a successful match. Typically, the original description and/or the descriptions of the new system lacked sufficient requirements specifications able to obtain a reasonable match. After a few months, the software-description database project quietly disappeared.

Another form of reuse involves the use of commercial software products developed and sold to meet a specific need. Examples are Microsoft Word, "apps" for telephones, Omni Pro, Lotus Notes, VisiCalc, and so forth. Many years ago, software users avoided a purchased (commercial) software system, instead insisting on a tailor-made software system.

Frequently, the commercial system did not "match" the buyer's business methods. Years ago, companies did not want to change their procedures to match the available software process. Nowadays, most enterprises are more than willing to change their procedures to match the types of software that can be purchased.

3.8 Software quality.

This feature can be represented as:

(1) The totality of features and characteristics of a software product that affect its ability to satisfy given needs (e.g., to conform to specifications).

(2) A characteristic of software that determines the degree to which the software will meet the expectations of the customer.

(3) An attribute of software that affects its perceived value, for example *correctness, reliability, maintainability, portability, etc.*

(4) Software quality includes *fitness for purpose, reasonable cost, reliability,*

ease of use in relation to those who use it, design of maintenance, upgrade characteristics, and favorable comparison against reliable products [ACARD 1986].

3.9 Improving construction quality.

Numerous techniques exist to ensure the quality of code as it is constructed. The specific techniques selected depend on the nature of the software being constructed, as well as on the skill sets of the software engineers performing the construction.

> *To improve the quality of the product you must improve the quality of the process (the Number 1 rule of software engineering).*

Techniques that can applied for improving code quality are as follows [McConnell 2004]:

(1) *Institute a set of coding guidelines for the development of* **code** — The company coding standard is what developers are expected to follow within the company, whereas the coding standard for a programming language is what the programming language developers recommend all programmers should follow.

(2) **Institute code walkthroughs or formal inspections** — Code walkthroughs or code inspections are very effective peer reviews in which the peers of the programmer that wrote the programs walk through or thoroughly inspect the program looking for errors. Managers are not invited or wanted at a walkthrough or inspection meeting

(3) *Institute a verification and validation (V&V) process to verify code* — Verification and validation are independent procedures, which are used together for checking that a product, service, or system meets requirements and specifications that fulfill its intended purpose.

(4) **Use external audits when necessary** — In software engineering a periodic or specific purpose audit conducted by external (independent) qualified programmers. Its objective is to determine, among other things, whether the software documenting (1) is accurate and complete, (2) reflects the software requirements specification, and (3) is prepared in accordance with the company standards.

(5) **Develop a reward structure** — Use a reward structure that rewards error-free producers of code rather than the error-prone producers of code. Develop a series of rewards for effective, efficient, and error-free code, publicize the reward program, periodically evaluate the code (including documentation) after it has been placed under configuration control, and give credit when credit is warranted.

Construction quality activities are differentiated from other quality activities by their focus. Construction quality activities focus on code and on artifacts related to code, like small-scale designs, as opposed to other artifacts that are less directly connected to the code, such as requirements, high-level designs, and project plans [SWEBOK 2014].

Additional construction quality techniques used to improve construction quality are software quality, unit and integration testing, test-first development, code stepping, and outside and inside documentation.

(6) ***Test-first development*** — This concept suggests that writing test cases first will minimize the time between a defect's insertion into the code and the defect's discovery and removal [McConnell 2004].

(7) ***Code stepping*** — One of the most common debugger usage scenarios is called "code stepping." When debugging under this scheme, the developer codes at a lower level with a more concrete construction.

(8) **Unit development folder (UDF)** — A UDF was typically filed in a three-ring notebook containing all pertinent information for a software unit. (A unit can also be synonymous with a module or a work package.)

The UDF provides a level of management control over these low-level software units. Periodically, the project manager will visit a software unit and review the UDF. The manager's initials on the UDF indicate that the document has been reviewed and accepted as the status to date.

Today a UDF is more typically maintained on-line in an electronic file.

(9) ***Outside documentation*** — What has classically been known as "documentation" is actually outside documentation—e.g., *users' manuals, operator's manuals* (recently combined with users' manuals) and *maintenance manuals.* Documents external to the system tend to be more abstract. Two examples of outside documentation are *software requirements documents (SRD)*, and/or a *ConOps* written by potential users of the system.

(10) ***Inside documentation*** — Inside documentation can be one of two types—*online help manuals* or *commented code.* Commented code is composed of useful comments that might be valuable when attached to a code module in a computer program.

Suggestions for structuring useful comments include [McConnell 1993]:

- Summary of the code module.

- Description of the code's intent.

- Use commenting styles that do not break down or discourage modification.

- Make comments about undocumented features or error "work arounds."

- Do not use styles that are too fancy or are too difficult to maintain.

- A program design language (PDL) is useful in making comments.

- Avoid cryptic comments and the use of inappropriate words.

McConnell points out that commenting style should not break down or discourage modification. Also, do not use any style that is too fancy, annoying, or inappropriate for the task. McConnell suggests that the developer or coder include comments about undocumented features or work-arounds in the comments to prevent another developer or coder from attempting to fix a work-around later in the maintenance life cycle.

3.10 Integration.

A key activity during construction is the integration of separately constructed routines, classes, components, and subsystems. In addition, a particular software system may need to be integrated with other software or hardware systems.

Four typical integrations are *"big bang" integration, incremental integration, top-down integration,* and *a top-down vs. bottom-up approach:*

(1) ***Big bang integration*** — In *big bang integration*, all components, modules, and subsystems are combined and integrated at one time. This is done under the assumption that the quickest way to integrate the system is in one giant step, "so we can get home by dinner." It is called "big bang" integration because it frequently "blows up," i.e., fails to integrate.

If the systems fail to integrate, determining the cause becomes next to impossible. When testing hardware systems, this is called "the smoke test" because the idea is "let's put the power on and see where the smoke rises."

Smoke, of course, is typically caused by a short or a bad component within the system. Unfortunately, there is no smoke in software to indicate a failure.

(2) ***Incremental integration*** — *Incremental integration* can be top-down or bottom-up. Top-down completes the process a little bit at a time. Starting with the small core of the system, the developer will design, code, test, and debug some part, then integrate the new part with the core.

The new partial system is tested to ensure its operability. If the test is successful, proceed to repeat the earlier steps with another untested part. If the system does not work, the probable cause lies either within the new code or is embedded in the integration between the new code and the system core.

(3) **Top-down integration** — Figure 5 presents an illustration of *top-down integration* given to me by Dr. Winston Royce when I worked for the Lockheed Software Technology Center, Austin, TX, in the early 1980s. Begin with Element 1 then add Element 2. Test the two together. If the process passes the test, add Element 3 to the new core. If it passes, add Element 4 to the core and so forth until each system element has been integrated and tested.

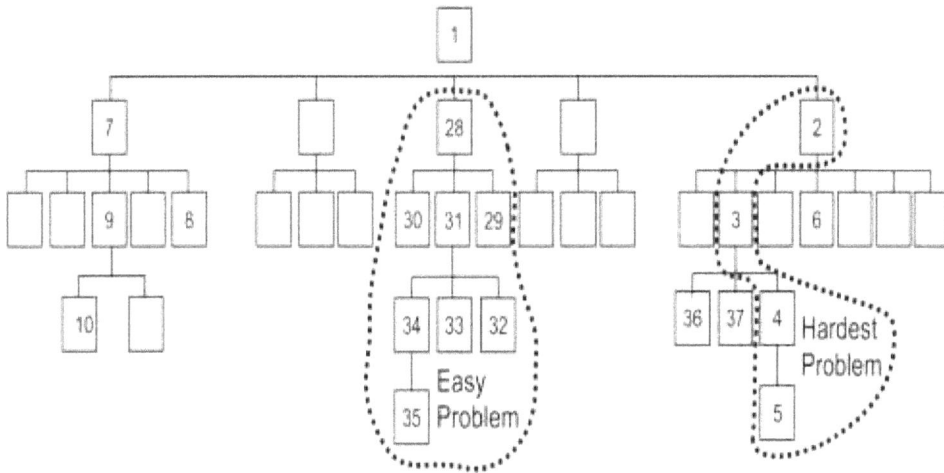

Figure 5: Top-down integration

A slight variation to this approach is to integrate several of the elements into the original core, making sure to include elements that can both read the data and output the data. Dr. Royce presented an argument for completing the hard part first to ensure the system could be built. There is an equally valid point for completing the easy part first. What is it?

The benefits of incremental integration are many. Errors are easy to locate because the last integration is probably at fault. Incremental integration provides early evidence that progress is being made. Planning is more accurate and components are tested more fully and frequently. The growing core is tested each time a new component is added. New test data should be added each time the system is tested to verify the new component. This improves the development and test schedule because work is being done in parallel.

(4) **Bottom-up integration** — In this application, the chart reflects top-down versus bottom-up integration. In *bottom-up integration*, lower code modules are built first, and then *drivers* are built to imitate the higher-level code modules. In *top-down integration*, the high-level codes are built first. *Stubs* are created that return a meaningful answer when

called by the higher-level module. Usually, top-down integration is considered superior to bottom-up integration.

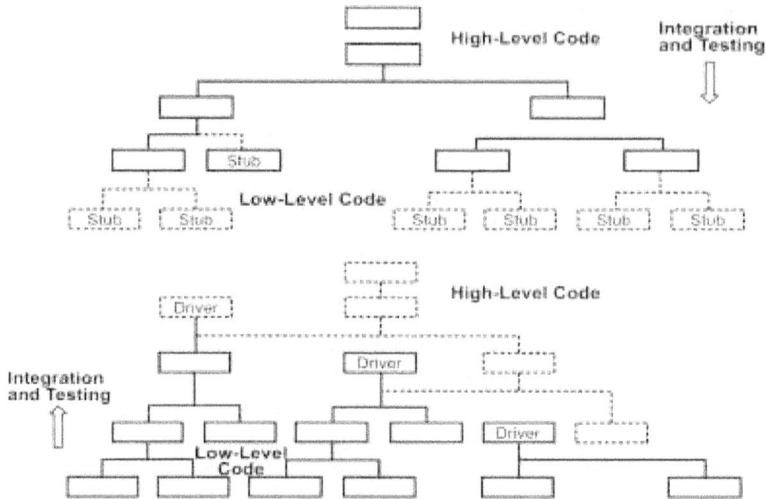

Figure 6: Illustration of top-down and bottom-up integration

4. Construction Technologies

"Technology" can be most broadly defined as the entities, both hardware and software, created by the application of mental and physical effort to achieve something of value. According to this usage, technology refers to tools and machines that may be used to solve real-world problems.

This broad definition may include simple tools such as a crowbar or wooden spoon, or complex machines such as a space station or particle accelerator. Tools and machines need not be physical; virtual technology, such as computer software and business methods, falls under this definition of construction technology.

The word "technology" can also be used to refer to a collection of techniques. In this context, technology is the current state of humanity's knowledge determining how to combine resources to produce desired products, solve problems, fulfill needs, or satisfy wants. It includes technical methods, skills, processes, techniques, tools, and raw materials. When combined with another term, such as "medical technology," "space technology," and "construction technology," technology refers to the state of the respective field's knowledge and tools. "State-of-the-art technology" refers to the highest level of technology available to humanity in any field [http://en.wikipedia.org/wiki/Technology].

4.1 API design and use.

An *application programming interface* (API) is a language and message format used by an application program to communicate with the operating system or control program such as a database management system. The API implies that either a program module is performing the operation, or a program must be linked into the existing program to perform the tasks [PC Magazine Encyclopedia].

4.2 Object-oriented run-time issues.

Object-oriented programming (OOP) is a programming paradigm that uses "objects" to design applications and computer programs. Programming techniques may include features such as *data abstraction, encapsulation, modularity, polymorphism, inheritance,* and *dynamic binding* [http://en.wikipedia.org/wiki/Object_oriented_programming].

(1) **Data abstraction** — *Abstraction* is the process by which data and programs are defined with a representation similar in form to its meaning (semantics), while hiding the implementation details.

(2) **Encapsulation** — *Encapsulation* refers to one of two related but distinct notions, and sometimes to their combination: (1) a language mechanism for restricting access to some of the object's components, and (2) a language construct that facilitates the bundling of data with the methods (or other functions) operating with the use of that data.

(3) **Modularity** — Also known as *modular programming, top-down design,* and *stepwise refinement*, modularity is a software design technique that increases the extent to which software is composed of separate, interchangeable components by breaking down program functions into *modules*, each of which usually accomplishes just one function.

(4) **Polymorphism** — *Polymorphism* is a programming language feature that allows values of different data types to be handled using a uniform interface. The concept of parametric polymorphism applies to both data types and functions. A function that can be applied to values of different types is known as a *polymorphic function.* A data type that can appear to be of a generalized type (e.g., a list with elements of arbitrary type) is designated a *polymorphic data type*, like the generalized type from which specializations are made [htpp://en.Wikipedia.org/wiki/Type_polymorphism].

(5) **Inheritance** — *Inheritance* is the capability of classes to inherit attributes from pre-existing classes (called *base classes, super classes, parent classes,* or *ancestor classes*). The resulting classes are known as *derived classes, subclasses,* or *child classes.*

By default, the subclass inherits all the attributes and properties of the super class, but these may be redefined in the subclass if desired. Inheritance may apply to objects as well as to classes. Note that this is a general definition, and individual languages may have specific inheritance

definitions that differ from the more general one [http://en.wikipedia
.org/wiki/Inheritance(object-orientedprogramming)].

(6) **Dynamic binding** — Dynamic binding or *late binding* can mean determining the exact implementation of a request based on both the request
(operation) name and the receiving object at run-time. Dynamic binding
often happens when invoking a derived class's member function using a
pointer to its base class. The implementation of the derived class will be
invoked instead of the base class. This allows substituting a particular
implementation using the same interface and enables polymorphism
[http://www.ask.com/9binding(computer_science)].

4.3 Parameterization and generics.

A *parameter* is a special kind of variable, used in a subroutine to refer to one of
the pieces of data provided as input to the subroutine. These pieces of data are
called *arguments*. An ordered list of parameters is usually included in the definition of a subroutine, so that each time the subroutine is called, its arguments for
that call can be assigned to the corresponding parameters. The term "argument"
is often (incorrectly) used in place of "parameter" [http://en.Wikipedia
.org/wiki/Parameter(computerprogramming)].

Two types of parameters are frequently used: dependent and independent variables. The *independent variable* is typically the variable representing the value
being manipulated or changed, and the *dependent variable* is the observed result
of the independent variable being manipulated. In Boehm's equations on software costs, the size of the computer program is the independent variable and
the cost is the dependent variable [Boehm 1981].

Generic programming is a style of computer programming in which algorithms
are written in terms of *to-be-specified-later* types that are then *instantiated*
(transformed) when needed for specific types provided as parameters. This approach, pioneered by the Ada system in 1983, permits writing common functions or types that differ only in the set of types on which they operate when
used, thus reducing duplication. Software entities created using generic programming are known as *generics* [http://en.wikipedia.org/wiki/Generic
_programming].

4.4 State-based and table-driven construction techniques.

State-based construction techniques are most commonly represented by *state
diagrams,* i.e., *state transition diagrams* (STD). A state diagram is a directed
graph in which each vertex represents a state and each edge represents a transition between two states [https://en.wikipedia.org/wiki/Finite-state_machine].

*Considered as a state machine (see Figure 7), a turnstile has two states:
locked and unlocked.*

Current state	Input	Next State	Output
Locked	Coin	Unlocked	Unlock turnstile so a customer can push through
	Push	Locked	None
Unlocked	Coin	Unlocked	None
	Push	Locked	When customer has pushed through, lock turnstile

Figure 7: State transition table

There are two inputs that affect the state: (1) putting a coin in the slot (coin) and (2) pushing the arm (push). In the locked state, pushing on the arm has no effect; no matter how many times the input push is given, it stays in the locked state. Putting a coin in—that is, giving the machine a coin input—shifts the state from locked to unlocked. In the unlocked state, putting additional coins in has no effect; that is, giving additional coin inputs does not change the state. However, a customer pushing through the arms shifts the state back to lock.

The turnstile state machine can be represented by a state transition table, showing for each state the new state and the output (action) resulting from each input.

The model operation of an FSM begins from one of the states (called a "start state"), proceeds through transitions depending on input to different states, and can end in any of those available; however, only a certain set of states mark a successful flow of operation (called *accept states*) [http://en.wikipedia.org/wiki /Finite-state_machine].

4.5 Error, exception handling, and fault tolerance.

Exceptional conditions are *activities* that occur in a system that are not expected or are not a part of normal system operation. *Exception handling* is the process of responding to these unusual occurrences. *Errors* are SW mistakes resulting in incorrect system operations. *Fault tolerance* is the property that enables a system to continue operating properly in the event of the failure of some of its components. When the system handles these exceptional conditions improperly, failures and system failures may occur.

4.6 Assertions, design by contract, and defensive programming.

These three programming technologies are defined as:

(1) **Assertions** — An *assertion* is a predicate (e.g., a true–false statement) placed in a program to indicate that the developer *thinks* that the predicate is always true [http://en.wikipedia.org/wiki/Assertion_(computing)].

(2) **Design by contract**™ — *Design by contract* or *programming by contract* is an approach used to design computer software. It prescribes that software designers should define formal, precise, and verifiable interface specifications for software components, which extend the ordinary definition of abstract data types with *preconditions, post-conditions,* and *invariants.*

These specifications are referred to as "contracts" in accordance with a conceptual metaphor including the conditions and obligations of business contracts. The term was coined by Bertrand Meyer in connection with his design of the Eiffel programming language [http://en.wikipedia.org/wiki/Design_by_contract].

(3) **Defensive programming** — *Defensive programming* is a form of defensive design intended to ensure the continuing function of a piece of software in spite of unforeseeable usage of the software. The idea can be viewed as reducing or eliminating the prospect of Finagle's law[2] from taking effect. Defensive programming techniques are used especially when a piece of software could be misused, whether mischievously or inadvertently, resulting in a potential catastrophic effect [http://en.wikipedia.org/wiki/Defensive_progaming].

4.7 Executable models.

Executable models are software products such as software requirements specifications and software design descriptions that can be run on a computer to produce an answer that satisfies the requirements or design. A simple executable model tool is a compiler that can convert source code to executable code.

4.8 Run-time configuration.

Using run-time configuration allows the developer to create and delete data services, adapters, and destinations, even after the server has been started.

There are many reasons why the developer might want to create components dynamically. For example, consider the following use cases:

---

2. Finagle's law: Anything that can go wrong, will go wrong, and at the worst possible moment.

(1) A separate destination for each office that uses an application is requested. Instead of manually creating destinations in the configuration files, they should be created dynamically based on information contained in a database.

(2) The developer wants to dynamically create, delete, or modify destinations in response to a user input.

There are two primary methods to perform dynamic configuration. The first is to use a *custom bootstrap service class*. This is the preferred practice to perform dynamic configuration. The second method is to *call a remote object* that performs dynamic configuration.

4.9 Internationalization and localization.

In computing, internationalization and localization are means of adapting computer software to the different languages, regional differences, and technical requirements of a target market. *Internationalization* is the process of designing a software application so that it can be adapted to various languages and regions without engineering changes.

Localization is the process of adapting internationalized software for a specific region or language by adding locale-specific components and translating the text. Some companies use the term "globalization" for the combination of internationalization and localization [http://en.wikipedia.org/wiki/Internationalization_and_localization].

This concept is also known as *national language support* or *native language support (NLS)*.

4.10 Grammar-based input processing (parsing).

To *parse* is to break down a statement into its component parts of speech with an explanation of the form, function, and syntactical relationship of each part.

4.11 Concurrency primitives.

Concurrency is a property of systems in which several computations are executing simultaneously and potentially interacting with each other. The computations may be executing on multiple cores in the same chip, on the same processor, or on physically separated processors.

Concurrency primitives are a series of processes that can be used to develop a computer program. Examples are semaphores and monitors.

(1) *Semaphores* — A *semaphore* is a protected variable or abstract data type that provides a simple but useful abstraction for controlling access by multiple processes to a common resource in a parallel programming environment.

A useful way to think of a semaphore is the record of the number of units belonging to a particular resource that are available to be coupled with operations in order to safely (i.e., without race conditions) adjust that record as units are required or become free, and if necessary, wait until a unit of the resource becomes available.

A semaphore is a useful tool in the prevention of race conditions and deadlocks; however, its use is by no means a guarantee that a program is free from these problems. Semaphores that allow an arbitrary resource count are called *counting semaphores,* while semaphores that are restricted to the values 0 and 1 (or locked/unlocked, unavailable/available) are called *binary semaphores* [http://en.wikipedia.org wiki/Semaphore_(programming)].

(2) **Monitors** — In concurrent programming, a *monitor* is an object or module intended to be used safely by more than one program thread. The defining characteristic of a monitor is that its methods are executed with mutual exclusion. That is, at each point in time, at most one thread may be executing a method. This mutual exclusion greatly simplifies reasoning about the implementation of monitors, compared with code that may be executed in parallel.

Monitors also provide a mechanism for threads to relinquish temporary exclusive access while waiting for a condition to be met before regaining exclusive access and resuming the task. Monitors also have a mechanism for signaling other threads that such conditions have been met [http://en.wikipedia.org/wiki/Monitor_(synchronization)].

4.12 Middleware (components and containers).

Middleware is used to describe a broad array of tools and data that help applications use networked resources and services. Some tools, such as authentication and directories, are present in all categorizations. Other services, such as cost scheduling of networked resources, secure multicast, object brokering, and messaging are the major middleware interests of particular communities, such as scientific researchers or business systems vendors. One definition that reflects this breadth of meaning is: "Middleware is the intersection of the stuff that network engineers don't want to do with the stuff that application developers also don't want to do" [http://middleware.internet2.edu/overview/middleware-faq.html].

(1) **Components** — *Components* of middleware are an executable unit of functionality. One can buy or download components, deploy them and they work. They are the equivalent of a software black box.

(2) **Containers** — *Containers* are used in application servers to plug components into application servers [Internet Encyclopedia 2003, p. 611].

4.13 Construction methods for distributed software.

The following twelve steps can be used to initiate construction approaches for developing a customer-oriented distributed software system (DSS) [Silberschatz, Galvin & Gagne 2008]:

(1) Draft a requirements specification for the DSS.

(2) Establish, with the potential user, a draft front-end interface.

(3) Inventory the technical capabilities of the software engineering development team.

(4) Schedule training gaps in the technical knowledge of the assigned software engineers.

(5) Estimate the schedule and cost for developing the system.

(6) Inventory the existing system to determine what part of the system can be reused.

(7) Establish the degree of authority that the project management, the user, and sponsor have over the project.

(8) Develop a prototype system. Reanalyze the requirements and user interface.

(9) Look again at the requirements. Are they realistic?

(10) Is the budget realistic?

(11) Do the benefits outweigh the costs and potential problems?

(12) Start the project.

4.14 Constructing heterogeneous systems (hardware and software).

In information technology, *heterogeneity* means a network comprising different types of computers, with vastly differing memory sizes, processing power, and even basic underlying architecture, or a data resource with multiple format types [http://en.wikipedia.org/wiki/Homogeneity_and_heterogeneity].

4.15 Hardware-software co-design.

Current methods for designing embedded systems require hardware and software to be specified and designed separately. A specification, often incomplete and written in non-formal language, is developed and sent to the hardware and software engineers. HW/SW partition is decided a priori and is adhered to as much as possible because any changes in this partition may necessitate extensive redesign. Designers often strive to make everything fit in the final software product and therefore off-load only a few design parts to hardware in order to meet timing constraints [Pederson 2011].

Lockheed Martin [2006] defines "codesign" as a simultaneous consideration of hardware and software within the design process, consisting of the "co-development and co-verification of hardware and software through the use of simulation and/or emulation." Co-design includes the following four components [Assimakopoulos 1998]:

(1) *Co-specification* — In which the roles of software and hardware in implementing system functionality are considered and, based on the evaluation, the implementation is assigned to either of the two.

(2) *Co-development* — During which the software, hardware, and interfaces are developed together.

(3) *Co-verification* — To further optimize and refine the SW/HW partitioning (i.e., to aid design-space exploration).

(4) *Co-management* — That covers coordination, project management, requirements management, and configuration management throughout system specification, development, and verification.

4.16 Performance analysis.

Performance analysis, commonly known as *profiling*, is the investigation of a program's behavior using information gathered as the program executes. Its goal is to determine which sections of a program to optimize.

A *profiler* is a *performance analysis tool* that measures the behavior of a program as it executes, particularly the frequency and duration of function calls. Performance analysis tools have existed from at least the early 1970s. Profilers may be classified according to their output types or their methods for data gathering.

4.17 Code tuning.

Tuning (or *code tuning*) is the practice of modifying correct code to make it run more efficiently. Tuning refers to small-scale changes that affect a single class, a single routine, or more commonly, a few lines of code. Tuning does not refer to large-scale design changes or other higher-level means of improving performance. There is an understanding that code tuning can make dramatic improvements at each level.

Jon Bentley [1982] cites an argument that in some systems the improvements at each level can be multiplied by a factor of 10. Because a 10-fold improvement can be achieved in each of six levels, there exists an implied potential performance improvement of a million-fold. Although such a multiplication of improvements requires a program in which gains at one level are independent of gains at other levels, the potential is inspiring.

However, code tuning is not an effective way to improve performance: program architecture, class design, and algorithm selection usually produce improvements that are more dramatic. Nor is code tuning the easiest way to improve performance. Buying new hardware or a compiler with a better optimizer is easier and maybe cheaper. Moreover, it is the least effective way to improve performance—it takes more time to hand-tune code initially, and hand-tuned code is harder to maintain later.

4.18 Platform standards.

A *computing platform* is a type of hardware architecture and software framework (including application frameworks) that allows software to run. Typical platforms include a computer's architecture, operating system, programming language, and related user interface (and may also include a run-time library or a graphical user interface).

A platform is a crucial element in software development. A platform might be simply defined as a place to launch software. The platform provider enters into an agreement with the software developer stating that logic code will interpret consistently as long as the platform is running on top of other platforms. *Logic code* includes byte code, source code, and machine code. Using logic code actually means that execution of the program is not restricted by the type of operating system provided. Logic code has largely replaced machine-independent languages [http://en.wikipedia.org/wiki/Computing_platform].

POSIX (Portable Operating System Interface) is a family of standards, specified by the IEEE, to clarify and make uniform the application programming interfaces (and ancillary issues, such as command line shell utilities) provided by Unix-like operating systems. When programs are written to rely on POSIX standards, they can be easily ported among a large family of UNIX derivatives [Martelli 2009].

Java refers to a number of software products and specifications developed by Sun Microsystems, a subsidiary of Oracle Corporation, that together provide a system for developing application software and deploying it in a *cross-platform environment*. For software to be considered cross-platform, it must be able to function on more than one computer architecture or operating system. Java is used in a variety of computing platforms, from embedded devices and mobile phones to enterprise servers and supercomputers.

Java platform is the collective name for a bundle of related programs produced by Sun that allow for the development and running of programs written in the Java programming language. The Java platform is not specific to any one processor or operating system, but is rather an execution engine (called a *virtual machine*). A compiler with a set of libraries implemented for various hardware and operating systems enables Java programs to run identically on all Java platforms. *Note: There is a distinction between Java as a language and Java as a platform* [http://en.Wikipedia.org/wiki/Java_(software_platform)].

4.19 Test-first programming.

Don Wells, a consultant for Extreme Programming, reports, "When you *create your tests first*, before the code, you will find it much easier and faster to create your code." The combined time required to create a unit test and create code to pass the test is approximately the same amount of time as coding the test from the beginning. However, if the unit tests are available when starting coding, they do not need to be created after the code is completed, saving time at this stage and significant time later during development. Creating a unit test helps a developer to consider what tasks need to be accomplished[http://www.extreme pro graming.org /rules/test first.html].

5. Construction Tools

A *programming tool* or *software development tool* (i.e., a *construction tool*) is a program or application that software developers use to create, debug, maintain, or otherwise support other programs and applications.

5.1 Development environments.

A *software development environment* or *integrated development environment (IDE)* is comprised of the entire environment (applications, servers, and network) that provides comprehensive facilities to computer programmers for software development.

Typically, an IDE is dedicated to a specific programming language, allowing a feature set that most closely matches the programming paradigms of the language. IDEs typically present a single program in which all development is created and designed. In addition to basic code editing functions, modern IDEs often offer other features like compilation and error detection from within the editor, integration with source code control, build/test/debugging tools, compressed or outlined views of programs, automated code transforms, and support for refactoring.

The aim is to abstract the configuration necessary to piece together command-line utilities in a cohesive unit, which theoretically reduces the time required to learn a language and increases developer productivity [http://en.wikipedia.org/wiki/Software_development_environment].

(1) ***Compilation and error detection from within the editor*** — Even experienced programmers make mistakes, so they appreciate any help a compiler can provide in identifying the mistakes. Because novice programmers may make frequent mistakes and may not understand the programming language very well, they need clear, precise, and jargon-free error reports. In a learning environment, the main function of a compiler is to report errors in source programs.

(2) ***Support for refactoring*** — *Code refactoring* is the process of restructuring existing computer code—changing the factoring—without changing its external behavior. Refactoring improves nonfunctional attributes of the software. Advantages include improved code readability and reduced

complexity to improve source code maintainability, and creation of a more expressive internal architecture or object model to improve extensibility [http://en.wikipedia.org/wiki/Code_refactoring].

5.2 GUI builders.

A *graphical user interface (GUI)* builder, also known as a *GUI designer*, is a software development tool that simplifies the creation of GUIs by allowing the designer to arrange widget controls using a drag-and-drop WYSIWYG (What You See Is What You Get) editor. Without a GUI builder, a GUI must be built by manually specifying each widget's parameters in code.

A *widget* (or *control*) is an element of a GUI that displays an information arrangement changeable by the user, such as a window or a text box [http://en.wikipedia.org/wiki/Graphical_user_interface_builder].

User interfaces are commonly programmed using an event-driven architecture, so GUI builders also simplify event-driven code. This supporting code connects widgets with both outgoing and incoming events that trigger the functions providing the application logic.

5.3 Unit testing tools.

Unit testing is a method by which individual units of source code are tested to determine if they are fit for use. A *unit* is the smallest testable part of an application. In procedural programming, a unit may be an individual function or procedure. Unit tests are normally created by the unit programmers.

White box testing (a.k.a. *clear box testing, glass box testing, transparent box testing,* or *structural testing*) is a method of testing software that tests internal structures or workings of an application as opposed to its functionality (i.e., black box testing). In white box testing, internal perspectives of the system, as well as programming skills, are required and used to design test cases. The *white box tester* chooses inputs to exercise paths through the code and to determine the appropriate outputs [http://en.wikipedia.org/wiki/Whitebox_testing].

Ideally, each test case is independent from the others: substitutes like *method stubs, mock objects, fakes,* and *test harnesses* can be used to assist in testing a module in isolation. Unit tests are typically written and run by the software developers who developed the code to ensure that code meets its design and performs as intended. Its implementation can vary from manual to formalized as part of build automation.

5.4 Profiling, performance analysis, and program slicing tools.

(1) *Software profiling* (or simply *profiling*), a form of *dynamic program analysis* (as opposed to static code analysis), is the investigation of a program's behavior using information gathered as the program executes. The usual purpose of this analysis is to determine what sections of a

program to optimize, e.g., to increase its overall speed, decrease its memory requirement, or sometimes both optimizations.

A *code profiler* is a performance analysis tool that most commonly measures only the frequency and duration of function calls. There are other specific types of profilers (e.g., memory profilers) in addition to more comprehensive profilers capable of gathering extensive performance data [http://en.wikipedia.org/wiki/Software_profiling].

(2) *Performance analysis* involves gathering formal and informal data to help customers and sponsors define and achieve their performance goals. Performance analysis uncovers several perspectives pertaining to a problem or opportunity, determining all drivers or barriers to successful performance, and proposing a solution system based on discoveries.

(3) *Program slicing* is the computation of the set of program statements, or the program slice, that may affect the values at some point of interest—a slicing criterion. Program slicing can be used in debugging to easily locate sources of errors. Other applications of slicing include software maintenance, optimization, program analysis, and information flow control [http://en.wikipedia.org/wiki/Program_slicing].

REFERENCES

Additional information explaining SWC knowledge areas can be found in the following documents:

- **[ACARD 1986]** "A Vital Key to UK Competitiveness," ACARD (U.K.), 1986.

- **[Assimakopoulos 1998]** N.A. Assimakopoulos, "Systemic Industrial Management of HW/SW Co-Design," *Journal of High Technology Management Research*, Vol. 9, no. 2, 1998, pp. 271–284.

- **[Bentley 1982]** Jon Bentley, *Writing Efficient Programs*, Prentice-Hall, Englewood Cliffs, NJ, 1982.

- **[Bidgoli 2003]** Hossein Bidgoli (ed.), *The Internet Encyclopedia*, Vol. 1, Wiley, Hoboken, NJ, 2003, p. 611.

- **[Boehm 1981]** Barry W. Boehm. *Software Engineering Economics*, Prentice-Hall, Englewood Cliffs, NJ, 1981.

- **[Clements, et al. 2002]** Paul Clements, Felix Bachmann, Len Bass, David Garlan, Paulo Merson, James Ivers, Reed Little, Robert Nord, and Judith Stafford, *Documenting Software Architectures: Views and Beyond*, Pearson, Boston, 2002. (Recommended as a reference for the IEEE PSEM Certification exam by the IEEE Computer Society.)

- **[IEEE Std. 610.12-1990]** *IEEE Standard Glossary of Software Engineering Terminology*. (Revision and redesignation of IEEE Standard 729-1983), IEEE, New York, 1990.

- **[IEEE Std. 1028-2008]** IEEE *Standard for Software Reviews,* IEEE, New York, 2008.

- **[Ingrassia 1976]** F.S. Ingrassia, "The Unit Development Folder (UDF): An Effective Management Tool for Software Development," *TRW Technical Report*, TRW-SS-76-11, 1976.

- **[Internet Encyclopedia 2003]** Hossein Bidgoli, *The Internet Encyclopedia*, Vol. 1, John Wiley, Hoboken, NJ, 2003.

- **[Lockheed Martin 2006]** Lockheed Martin, "Hardware/Software Codesign" 2006. [http://www.atl.lmco.com/projects/rassp/RASSP_legacy/app notes/HWSW/APNOTE_HWSW_INDEX.HTM]

- **[Martelli 2009]** Alex Martelli, "Answers to questions on stack overflow", 2009. [http://stackoverflow.com/questions/1780599/i-never-really-understood-what-is-posix]

- **[McConnell 1993]** Steve McConnell, *Code Complete*, Microsoft Press, Redmond, WA, 1993.

- **[McConnell 2004]** Steve McConnell, *Code Complete*, 2nd ed., revised, Microsoft Press, Redmond, WA, 2004. (Recommended as a reference for the PSEM Certification exam by the IEEE Computer Society.)

- **[Null and Lobur 2006]** Linda Null and Julia Lobur, *The Essentials of Computer Organization and Architecture*, 2nd ed., Jones & Bartlett, 2006. Chapters 1-4, 9-12, and Sections 8.1-8.4, 8.6, and 8.7. (Recommended as a reference for the PSEM Certification exam by the IEEE Computer Society.)

- **[Parnas 1972]** David L. Parnas, "On the Criteria to Be Used in Decomposing Systems into Modules," *Communications of the ACM*, Vol., 15, no. 12, 1972, pp. 1053–1058.

- **[PC Magazine Encyclopedia 1986]** *PC Magazine Encyclopedia*, Ziff Davis, New York, 1996.

- **[Pederson 2011]** Donald O. Pederson, "A Framework for Hardware-Software Co-Design of Embedded Systems," *UCB Electronic Systems Design Publications*, University of California, Berkeley, CA, 2011.

- **[Royce 1970]** W.W. Royce, "Managing the Development of Large Software Systems: Concepts and Techniques," *1970 WESCON Technical Papers*, Vol. 14, Western Electronic Show and Convention, Los Angeles, Aug. 25-28, 1970.

- **[Sakharov 2005]** "Finite State Machines," 2005. [http://sakharov.net /fsmtutorial.html]

- **[Silberschatz, Galvin and Gagne 2008]** J. Abraham Silberschatz, Peter Baer Galvin, and Greg Gagne, *Operating System Concepts*, 8th ed., John Wiley, Hoboken, NJ, 2008, Chapters 3-6, 16, 18. (Recommended as a reference for the PSEM Certification exam by the IEEE Computer Society.)

- **[SWEBOK 2004]** Guide to the *Software Engineering Body of Knowledge (SWEBOK)* IEEE, New York, 2004.

- **[SWEBOK 2014]** Guide to the *Software Engineering Body of Knowledge (SWEBOK)*, IEEE, New York, 2014.

- **[Wikipedia]** Wikipedia is a free, web-based encyclopedia enabling multiple users to freely add and edit online content. Definitions cited on Wikipedia and their related sources have been verified by the authors and other peer reviewers.

Chapter 2

Software Engineering Construction[3]

Steven McConnell
Construx Software

Developing computer software can be a complicated process, and in the last 25 years, researchers have identified numerous distinct activities that are part of software development.

If you have worked on informal projects, you might think that our list represents a lot of red tape. If you have worked on projects that are too formal, you know that our list represents a lot of ted tape. It is hard to strike a balance between too little and too much formality.

If you have taught yourself to program or worked mainly on informal projects, you might not have made distinctions among the many activities that are involved with creating a software product. Mentally, you might have grouped all of these activities together as "programming."

If you work on informal projects, the main activity you think of when you think about creating software is probably the activity the researchers refer to as "construction."

1. What is Software Construction?

This intuitive notion of "construction" is accurate, but it suffers from a lack of perspective. Putting construction in its context with other activities helps to keep the focus on the right tasks during construction and appropriately emphasizes important non-construction activities.

Figure 1 represents my vision of software engineering construction (SEC). The dark circle represents the core of SEC. The "funny looking" rectangles represent the activities of software engineering construction. The degrees in which the rectangles overlap the core represent the percentage of the activities included in the SEC core.

Construction is mostly coding and debugging but also involves limited requirements, detailed design, construction planning, unit testing, integration, integration testing, and other activities. If this were a book about all aspects of software development, it would feature nicely balanced discussions of all activities involved in the development process. This is a paper about software construction and as a result, it places a lopsided emphasis on construction and only touches on related topics.

---

3. Extract from McConnell's book, *Code Complete: A practical handbook of software engineering construction*, 2nd ed., Microsoft Press, Redmond, WA, 2004. Used with permission of Microsoft Press and the author. (This book is recommended as a reference for the IEEE PSEM Certificate exam by the IEEE Computer Society.)

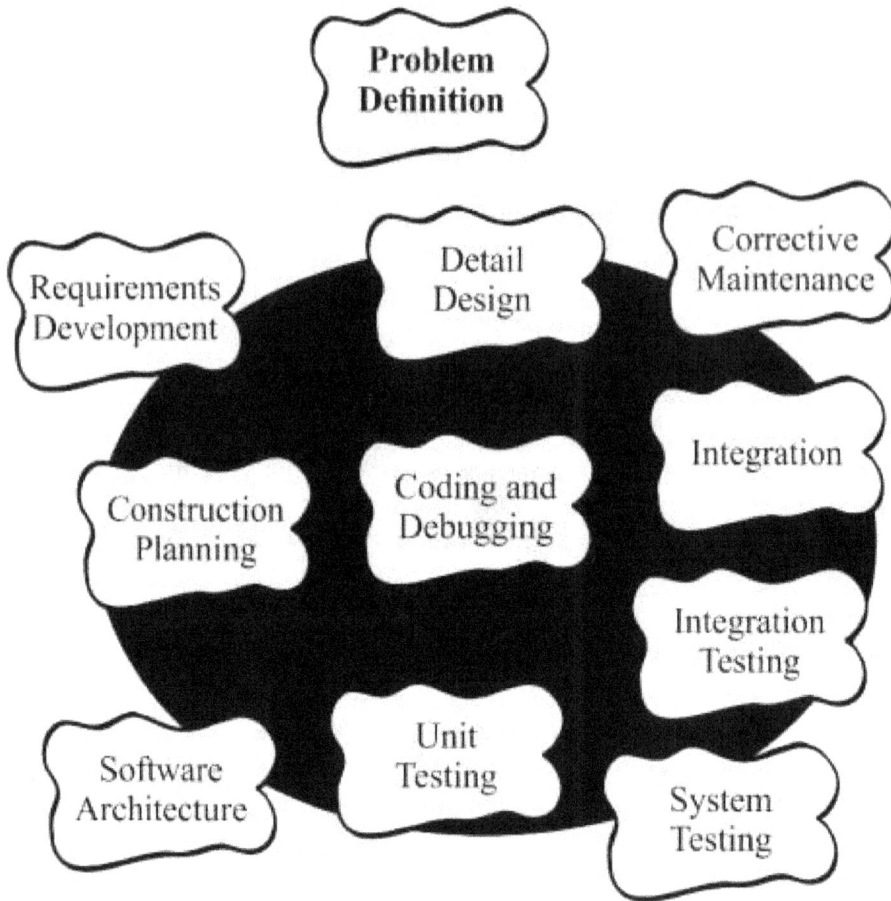

Figure 1: Software construction

Here are some of the specific tasks involving construction:

- Verifying that the groundwork has been laid so that construction can proceed successfully.
- Determining how your code will be tested.
- Designing and writing classes and routines.
- Creating and naming variables and named constants.
- Selecting control structures and organizing blocks of statements.
- Unit testing, integration testing and debugging your own code.
- Reviewing other team members' low-level designs and code and having them review yours.
- Polishing code by carefully formatting and commenting it.

- Integrating software components that were created separately.
- Tuning code to make it faster and use fewer resources.

With so many activities at work in construction, you might say, "OK, Jack, what activities are *not* parts of construction?" That is a fair question. Important non-construction activities include management, requirements development, software architecture, user-interface design, system testing, and maintenance. Each of these activities affects the ultimate success of a project as much as construction—at least the success of any project that calls for more than one or two software developers and lasts longer than a few weeks. Well-written books are available for each of these subjects and are listed in the "Additional Resources" section of my book *Code Complete*.

2. Why is Software Construction Important?

You probably agree that improving software quality and developer productivity are important. Many of today's most exciting projects use software extensively. The internet, movie special effects, medical life-support systems, space programs, aeronautics, high-speed financial analysis, and scientific research are a few examples. These projects and more conventional projects can all benefit from improved practices; many of the fundamentals are the same.

If you agree that improving software development is important in general, the question for you as a reader of this chapter becomes—Why is construction an important focus? The following five reasons explain the value of construction:

(1) ***Construction is a large part of software development*** —Depending on the size of the project; *construction* typically takes 30 to 80 percent of the total time spent on a project. Any component that takes up that much project time is bound to affect the success of the project.

(2) ***Construction is the central activity in software development*** — Requirements and architecture are done before construction to enable construction to be effective. System testing (in the strict sense of independent testing) is performed after construction to verify that construction has been designed correctly. *Construction* is at the center of the software development process.

(3) ***Why focus on construction?*** — The individual programmer's productivity can improve enormously. In a classic study undertaken by Sackman, Erikson and Grant [1968] the researchers showed that the productivity of individual programmers varied by a factor of 10 to 20 during construction. Since publication, their results have been confirmed by numerous other studies [Curtis 1981, Mills 1983, Curtis et al. 1986, Card 1987, Valet and McGarry 1989, DeMarco and Lister 1999, Boehm, et al. 2000]. My book, *Code Complete*, can help all programmers to learn techniques that are being used by the best programmers.

(4) ***Construction's product*** — The source code is often the only accurate description of the software. In many projects, the only documentation available to programmers is the code itself. Requirements specifications and design documents can become outdated, but the source code is always up to date. Consequently, it is imperative that the source code be of the highest possible quality. Consistent application of techniques for source-code improvement makes the difference between a "Rube Goldberg" contraption and a detailed, correct, and therefore, informative program. Such techniques are most effectively applied during construction.

(5) ***Construction is the only activity that is guaranteed to be done*** — The ideal software project goes through careful requirements development and architectural design before construction begins. The ideal project undergoes comprehensive and statistically controlled system testing after construction. Imperfect, real-world projects, however, often skip requirements and design and move directly to construction. Project managers will often drop testing because there are too many errors to fix and they have run out of time. Nevertheless, no matter how rushed or how poorly planned the project is, construction cannot be dropped. This is where the rubber meets the road. Improving construction is thus a way of improving any software-development effort, no matter how short.

3. Software Construction: Building Software

The image of "building" software is more useful than that of "writing" or "growing" software. It is compatible with the idea of software accreditation and provides guidance that is more detailed. Building software implies various stages of planning, preparation, and execution that vary in kind and degree depending on what is being built. When exploring this metaphor, many other parallels are found.

Building a four-foot tower requires a steady hand, a level surface, and 10 undamaged beer cans. Building a tower 100 times that size does not merely require 100 times as many beer cans. It requires a different plan and construction altogether.

If you are building a simple structure—a doghouse, say—you can drive to the lumber store and buy some wood and nails. By the end of the afternoon, Fido will have a new house. If you forget to provide for a door, or make some other mistake, it is not a big problem; you can fix it or even start over from the beginning. All you have wasted is part of an afternoon. This loose approach is appropriate for small software projects, too. If the wrong design for 1000 lines of code is used, one can refactor or start over completely without losing much.

If you are building a house, the building process is more complicated, and so are the consequences of poor design. First, you have to decide what kind of house you want to build—analogous in software development to problem definition.

Then you and an architect have to create a general design and get it approved. This is similar to software architectural design. You draw detailed blueprints and hire a contractor.

This in turn is similar to detailed software design. You prepare the building site, lay a foundation, frame the house, add siding and a roof, plumbing, and wiring. This is similar to SEC. When the house is nearly complete, the landscapers, painters and decorators arrive to add finishing touches to the property and the home you have built. This is similar to software optimization. Throughout the process, various inspectors come to check the site, foundation, framing, wiring, and other assessments. This is similar to software reviews and inspections.

Greater complexity and size imply greater consequences for both activities. When building a house, although materials are somewhat expensive, the main expense is labor. Ripping out a wall and moving it six inches is expensive not because of wasted nails but because of paying the labor for the extra time it takes to move the wall. The design must be as good as possible in order not to waste time fixing mistakes that could have been avoided. When building a software product, materials are even less expensive, but labor costs just as much. Changing a report format is just as expensive as moving a wall in a house—the main cost component in both cases is people (labor hours). What other parallels do the two activities share? When constructing a house, you will not try to build things that can be bought already built. You will buy a washer and dryer, dishwasher, refrigerator, and freezer. Unless you are a mechanical wizard, you will not consider building these yourself. Prefabricated cabinets, counters, windows, doors, and bathroom fixtures will also be purchased. When building a software system, the same approach is taken.

You will make extensive use of high-level language features rather than writing your own operating-system-level code. You might also use prebuilt libraries of container classes, scientific functions, user interface classes, and database-manipulation classes. It generally does not make sense to code components that can be bought ready-made.

If you are building a fancy house with first-class furnishings, however, you might have the cabinets custom-made. You might have a dishwasher, refrigerator and freezer built-in to look like the rest of the cabinets. You might have windows custom-made in unusual shapes and sizes. This customization has parallels in software development. If you are building a first-class software product, you might build your own scientific functions for better speed or accuracy. You might build your own container classes, user interface classes, and database classes to give the system a seamless, perfectly consistent look and feel.

Both building construction and SEC benefit from appropriate levels of planning. If you build software in the wrong order, it is hard to code, hard to test and hard to debug. It can take longer to complete, or the project can fall apart because everyone's work is too complex and, therefore, confusing when combined.

4. Select Major Construction Practices

Part of preparing for construction is deciding which of the many available good practices can be selected. Some projects use pair programming and test-first development, while others use solo development and formal inspections. Either combination of techniques can work well, depending on the specific circumstances of the project.

The following checklist summarizes the specific practices you should consciously decide to include or exclude during construction. Details of these practices are contained throughout *Code Complete*.

4.1 Coding.

(1) Have you defined how much design will be constructed up front and how much will be done at the keyboard while the code is being written?

(2) Have you defined coding conventions for names, comments and layout?

(3) Have you defined specific coding practices that are implied by the architecture, such as how error conditions will be handled, how security will be addressed, what conventions will be used for class interfaces, what standards will apply to reused code, how much to consider performance while coding, and so on?

(4) Have you identified your location on the technology wave and adjusted your approach to match? If necessary, have you identified how you will program *into* the language rather than being limited by programming *in* it?

4.2 Teamwork.

(1) Have you defined an integration procedure—that is, have you defined the specific steps a programmer must follow before delivering the code to the *program librarian*?

(2) Will programmers program in pairs, individually or some combination of the two?

4.3 Quality assurance.

(1) Will programmers write test cases for their code before writing the code itself?

(2) Will programmers write unit tests for their code regardless of whether they are written first or last?

(3) Will programmers step through the code in the debugger before they check it in?

(4) Will programmers integration-test their code before they check it in?

(5) Will programmers review or inspect each other's code?

4.4 Tools.

(1) Have you selected a revision control tool?

(2) Have you selected a language and language version or compiler version?

(3) Have you selected a framework such as J2EE or Microsoft.NET or explicitly decided not to use a framework?

(4) Have you decided whether to allow nonstandard language features?

(5) Have you identified and acquired other tools you will be using—editor, refactoring tool, debugger, test framework, syntax checker, and so forth?

5. Some of the Key Points of Software Construction

(1) Software construction is the central activity in software development; construction is the only activity that is guaranteed to happen on every project.

(2) The main activities in construction are detailed design, coding, debugging, integration, and developer testing (unit and integration testing).

(3) Another common term for construction is "programming."

(4) The quality of the construction substantially affects the quality of the software.

(5) In the final analysis, your understanding of software construction determines your level of programming, and that is the subject of my book—*Code Complete* [McConnell 2004].

The following list *summarizes* the specific practices you should consciously decide to include or exclude during construction.

- Requirements development.
- Construction planning.
- Software architecture or high-level design.
- Detailed design.
- Coding and debugging.
- Unit testing.
- Integration.
- Integration testing.
- System testing.
- Corrective maintenance.

REFERENCES

- **[Boehm, et al. 2000]** Barry Boehm, *Software Cost Estimation with Cocomo II*, Addison-Wesley, Boston, 2000.

- **[Card 1987]** David N. Card, "A Software Technology Evaluation Program". *Information and Software Technologies*, Vol. 29, no 6., July/August, 1987, pp. 291-300.

- **[Curtis 1981]** B. Curtis (ed.), "Tutorial: Human factors in software development." *Compsac 81*, IEEE Computer Society's Fifth International Computer Software and Applications Conference, Chicago, IL, November 16-20, 1981.

- **[Curtis, et al. 1986]** B. Curtis, E. Soloway. R. Brooks. J. Black, K. Ehrlich, and H.R. Ramsey, "Software Psychology: The need for an interdisciplinary program," *Proceedings of the IEEE*, Vol. 74 (8), pp. 1092–1106.

- **[DeMarco and Lister 1999]** Tom DeMarco and Timothy Lister, *Peoplewear: Productive Projects and Teams*, 2nd ed., Dorset House, New York, 1999.

- **[McConnell 2004]** Steve McConnell, *Code Complete: A Practical Handbook of Software Construction,* 2nd ed., Microsoft Press, Redmond, WA, 2004.

- **[Mills 1983]** Harlan D. Mills, *Software Productivity*, Little, Brown & Co., 1983.

- **[Sackman, Erikson and Grant 1968]** H. Sackman, W.J. Erikson, and E.E. Grand, "Exploratory experimental studies comparing online and offline programming performance." *Communications of the ACM,* Volume 11, issue 1, Jan. 1968, pp. 3-11.

[Valet and McGarry 1989] J. Valett and F.E. McGarry, "A Summary of Software Measurement Experiences in the Software Engineering Laboratories, *Journal of Systems and Software* Vol. 9, no. 2, February 1987, pp. 137-148.

Chapter 3
Software Construction:
Design, Write, and Test the Code

Mark Christensen, PhD,
Independent Consultant

1. Introduction

So the requirements have been approved, the architectural description is complete, the database has been partitioned, the subsystems identified, the top-level algorithms specified, and the user interface and dialogs have been prototyped. Now the fun begins: we get to write and test the code, or more precisely, more code, since we probably have been following an incremental development model. Regardless, we are about to enter into the activities that first enticed the majority of us into the business of software development in the first place.

Should it be a surprise that these same activities make up the bulk of the cost (that is, labor hours), or consume the bulk of the time (that is, schedule) of most software development projects? Well, surprise or not, it is a fact. Software construction involves a distribution of the total effort of a project by the activities associated with the detailed design, code, unit test of the product's custom components, and integration of those components, together with any off-the-shelf components, to make the final product.

1.1 Construction defined.

The activities of design, code, and unit test and integration have historically been collectively referred to as implementation. *Construction* refers to the major portion of these activities but excludes the architectural level of design [McConnell, 1993]. Included in this definition is the detailed planning needed to perform the technical tasks. It could be argued that the inclusion of detailed planning is not a major insight, since it always seems reasonable to assume that any major activity, such as implementation, would have to be planned.

Experience presents a different story. A good reason for emphasizing detailed planning is the fact, discussed above, that the construction activities make up fully 80 percent of the effort expended in a project. Such planning is especially called for in the detailed design, code and unit test, and integration and product test phases. In many cases, initial deployment is included in the product test but must be planned in a discrete manner, separate from the activities covered by simplistic life-cycle models. If the distinction is not made clear, the result can cause confusion for both customer and developer. This is especially true if the project is delivering incremental versions of the product, or developing and releasing a new version with enhanced features, while concurrently correcting problems in the earlier releases.

In a perfect world, there would be no overlaps. The product integrates without finding residual unit test defects. The system test finds no latent integration problems. The deployment proceeds smoothly and without event. The user finds no problems post-deployment. This never happens, and so we are confronted with the reality of necessary detailed planning.

Latent code and unit test problems are found during integration. System tests uncover a host of problems with the integrated product. The first time the product is deployed in the real world, a vast variety of unexpected situations arises during installation, which the product may or may not handle correctly. Once installed, other problems arise, either outright defects or extensions that the user earnestly desires.

When such overlaps occur, the construction activities are in full swing, with project expenditures at their greatest and the potential negative impacts to quality at their highest. This all raises the stakes, making the construction activities all the more important. These activities lay the foundation on which all later activities occur and all activities from the viewpoint of the original developer of a software unit are maintenance activities.

1.2 Current construction practices.

Features of software development today include the variety of products being built, the environments in which they must execute, and the methods and tools being used to build them. It is possible to identify at least four distinct types of software products and environments:

(1) *Classical information systems* — Such systems typically operate in controlled environments (classically a mainframe but also on a server), are operated by company staff who are familiar with the business functions and processes they automate, and can be very large and monolithic, often providing their functionality through special-purpose user interfaces and organization-specific (often proprietary) databases.

Systems may need to operate on a network but the execution environment is usually carefully monitored and controlled. Such systems are typically very large when taken together but may contain smaller subsystems. The construction activities are typically undertaken using developmental computers that are like (if not actually the same as) the computer systems on which the applications will eventually be executed.

(2) *Embedded systems* — Computers are embedded in a variety of products and devices today, from large-scale automotive, avionics, and factory-control systems, cameras, health-care devices large and small, toasters, ovens and refrigerators, and almost any other class of product. The con-

struction environment is usually very different from that on which the application will operate. The application environment (into which the software is embedded) may not have a user interface in the classical (screen or windows) sense.

(3) ***Desktop and local network applications*** — Examples of this class of application include stand-alone analysis, text or data manipulation, and communications tools. A specialized application to perform mechanical stress analysis is one example, while a stand-alone database is another. Such applications share many attributes with classical information systems, but usually involve a significant amount of local processing (at the desktop).

These applications may draw data across a local network. Construction activities are usually performed on the computer used to operate the application. Access to the network can be well understood (since it is local) and hence, will be largely under the control of the developing organization. Therefore, security is not usually a serious problem, assuming the organization's overall network is secure.

(4) ***Internet and wireless applications*** — This class of applications shares some of the attributes of each of the above three software product types and environments with the added complication that the communication vehicle (*the Net*) is not intrinsically secure. In the past, if an enterprise desired to make an application available to remote users (or to the public for that matter), the users would need explicit access to the enterprise computer system.

Another option was for the user's desktop (or local computer) to have a special front-end application (now called a client) that could mask the interface with the enterprise computer. The Net and Net browsers eliminate the need for such special interfaces, allowing much smaller organizations to offer computer-based facilities to their business partners.

It is also common for Net-based applications to download executable components onto the user's computer system without their knowledge. As a result, such applications share some of the attributes of embedded systems, in that the execution environment is not that in which the development occurs.

1.3 Constructive activities.

While each of these four classes of systems poses unique challenges and issues, the construction activities of building the product have the same focus. Following are seven shared characteristics:

(1) Evaluation of the requirements and architectural description to allow construction to begin on a known, firm foundation.

(2) Detailed construction planning for an efficient and effective construction effort.

(3) Elaboration of the design to the point where individual executable components are identified and specified.

(4) Detailed component design, sufficiently detailed so coding can begin.

(5) Coding and testing of the modules or components, which brings into being the components of the design.

(6) Transition to integration, allowing for an orderly, managed process of "assembling" the components to produce the product.

(7) Integrating and testing the product, preparing it for delivery.

The above processes are transparent to the fact that certain components may be purchased "off-the-shelf." Such components must be evaluated and "qualified" just as the coded components must be unit tested.

2. Selecting a Language and Tool Set

The factors that usually have the greatest impact on choosing a programming language generally fall into two categories: those that are intrinsic to the language and those that are artifacts of a particular implementation of the tool set. Some factors, however, are impacted by both. Examples of these are:

- ***The available development and execution environments*** — The environments (computer, operating system, and network) may restrict the choice of language directly (the operating system of the execution environment may interface poorly with the language at runtime, for example) or indirectly (there may be limited support of development tools for the language in the operational environment).

- ***Memory utilization*** — Compiling the same component (subroutine, function, object, or method) may produce object code of radically differing sizes, depending on which vendor's tool set is used. Data structures and objects (often depending on how data fields are packed into structures) may consume more or less memory, depending on the particulars of a language implementation.

  At the same time, the linguistic structure and runtime approach of each language (Basic, C, Ada, Pascal, FORTRAN, and Java) and its runtime environment and libraries determine the amount of memory required to produce a working product. These factors are themselves modified by the decisions made by the tool-set vendor.

  To a greater or lesser degree, such language-intrinsic and tool implementation-specific issues influence all criteria used to evaluate candidate languages. McConnell [McConnell, 1993] provides a table of "Best and Worst" languages for a variety of program types, which is excerpted below.

As McConnell notes, these are *general* statements about the languages. For example, some FORTRAN implementations are among the fastest available of any language, approaching that of Assembler and exceeding those of some C implementations. Likewise, many vendors of numerical packages now make them available in C++. In addition, most real applications combine attributes and features of multiple program types.

Thus, a 3D computer-aided design system will perform complex numerical computations on complex data types, often using dynamic memory, and operating in a real-time manner. Finally, the table does not include Internet or client-server applications, nor does it mention Java, many of whose implementations, for example, can execute in limited-memory environments. Nonetheless, the table is useful as a model for building an evaluation checklist for candidate languages and their implementations.

Table 1. The best and worst languages for particular kinds of programs [McConnell, 1993]

Program type	Best language	Worst language
Structured data	Ada, C/C++, Pascal	Assembler, Basic
Quick-and-dirty applications	Basic	Ada, Pascal, Assembler
Fast execution	Assembler, C/C++	Interpreted languages
Mathematical calculation	Fortran	Pascal
Easy-to-maintain	Pascal, Ada	C, Fortran
Dynamic memory use	Pascal, C/C++	Basic
Limited-memory environments	Basic, Assembler, C/C++	Fortran
Real-time program	Ada, Assembler, C/C++	Basic, Fortran
String manipulation	Basic, Pascal	C/C++

Many organizations will already have programming practices and procedures targeted towards one or more languages. Trained resource pools of employees familiar with these languages are often available, and may significantly influence the selection of language or tool set. Likewise, some languages are particularly well suited to certain "application frameworks" such as CORBA or COM. If the project or organization has selected such a framework, the choice of programming language may be automatic.

One approach to selecting a programming language and language tool-set vendor is to determine the available development and execution environments (the latter should be specified in the requirements; the former is often limited by company practice and finances). Next, determine what memory and computing power will be available to the application. Lastly, build a list of evaluation criteria using the above table as a starting point, augmenting it with other program attributes that are likely to influence the selection of the language and the tool set. The design approach may itself be one of these. For example, the Basic language is poorly suited to designs based on data flow, while a language such as Ada is better suited for this requirement.

It may be necessary to experiment with some of the language implementations to evaluate their expansion rates (bytes of memory per line of code, or fields of data structures). The expansion rate can then be applied to the line of code and data item count estimates to predict the memory utilization of the application. If one of the candidate languages/tool sets consumes more memory than is available (within a specified margin), it should be eliminated from consideration. Similarly, timing studies can be performed if a critical evaluation criterion exists. Such evaluations will provide "hands-on" experience with candidate languages and tools. They may in extreme cases, uncover "show stoppers" in the tool set, or other serious quality problems.

In addition to the programming language, there are many tools needed by the construction team. These include:

- Compilers, assemblers, and linkers.
- Vendor-supplied code libraries in source or object code format.
- Debuggers.
- Static code analyzers.
- Execution profilers.
- Testing tools, including test case recording and playback capabilities for regression testing.

In addition to the above, the construction team will also need a configuration management (CM) system to manage the detailed designs of the components, along with the code and test cases, and integration test plans. This is discussed in Section 5, "Detailed Construction Planning," but the other tools must at least be capable of coexisting with the CM system, if not actually integrating with it. The CM system should be in place as a matter of institutional practice and should be used to manage and stabilize the requirements and architectural description documents.

3. Evaluation of Requirements

It may seem strange that the first construction activity is an evaluation of the requirements and the architecture. After all, didn't we (the development team) just complete those tasks? Well, yes, but that was a different "team." In the first place, the focus of the team was on negotiating requirements with the customer and with satisfying those requirements in architecture. During the construction phase, a different mindset must be adopted: The team is about to start "cutting metal." In addition, the team has many more members during the construction phase than it had during earlier phases. So many more minds with differing viewpoints are brought to bear on the problem. Thus, a new team has been compiled.

By evaluating the requirements and the architectural description, a third and equally valid reason presents itself. The new, expanded team can become familiar with the intent (requirements) of the product and with the architecture into which their individual components will fit. The goals of the project (especially the effective and efficient application of labor resources) are better served by assigning the team an objective that produces a product (the evaluation report or memo) than by providing reading assignments or lecturing on the application under the title of "training."

Finally, by examining the requirements and the architecture, the construction team can also begin to plan the testing and integration of the product.

Earlier parts of this book have dealt extensively with the development, documentation and evaluation of requirements by developers and end users. As a practical matter, the viewpoint of the construction team is very different: It has to both build "the damn thing" (also known as the product), while simultaneously maintaining the requirements (which always evolve). Hence, the team's focus on the classical questions that arise in evaluating requirements is different.

For example, the construction team will often have little to add in the area of completeness ("Are all the user-requirements present?") beyond what *should* have been accomplished earlier. Likewise, the team often has a lot to say in the area of feasibility ("Can the user-requirements be satisfied?") or consistency ("Do some of the requirements conflict with others?). Why? Because the construction team is the one that has to make all the wishes and dreams become reality.

The construction team should *not* have to scurry around looking for, collecting, analyzing, and documenting requirements. If it has to, then so be it. But the team should clearly inform management that it is doing so. This work should have been completed in earlier phases of the project. If the project is following an iterative model, the construction team exercises care to ensure it is working within the scope of the current iteration, or developmental increment before sounding the alarm.

With that introduction, the topics that are often worth exploring during an evaluation [IEEE Standard 830-1998], [Christensen and Thayer, 2002] can be grouped into two broad categories: the properties that individual requirements should possess, and the properties that the requirements taken as a whole should possess. Individual requirements should be:

- Necessary.
- Traceable.
- Measurable.
- Unambiguous.
- Testable.

Taken together, the requirements should be:

- Ranked for importance (to the customer).
- Ranked for stability (or the likelihood of change).
- Modifiable (that is, the amount of change induced into one requirement by changing another).
- Complete.
- Consistent.

In practice however, these broad topics will be elaborated upon, especially in the area of completeness, where topics that are more detailed must be explored [McConnell, 1993], [Christensen and Thayer, 2002]. These detailed topics can be encompassed by three overarching principles that apply to the requirements individually and collectively: *completeness, consistency* and *conciseness.*

In the general area of *completeness*, the team should ask:

- Are functions requested by the user adequately described?
- Are externally observable or related sequences of functions specified?
- Are exclusion rules between the functions specified?
- Are response times for the functions specified?
- Are inputs, outputs and data needed to perform those functions identified?
- Are necessary assumptions about, and validity checks on, the inputs specified?
- Are formats of reports and screens specified?
- Are interfaces to other software and hardware systems and the communication mechanisms specified?

- Are necessary formulas and equations essential to the performance of the functions provided?

- Is the accuracy of numeric results specified?

- Are resource constraints (memory, storage and processor) specified?

- Are communication hand shaking, retry and error-handling protocols specified?

- Are responses to internally and externally generated error conditions specified?

- Are levels of security and reliability clearly described and specified?

- Are the requirements testable, both individually and, where it makes sense, together?

In the general area of *consistency*, are any requirements:

- In conflict with one or more requirements?

- Specified in more (and more excessive) detail than others?

In the general area of *conciseness* (minimalists) are the requirements:

- All necessary to the solution of the problem?

- Individually specified only to the degree necessary to describe the problem, not the solution?

At all times, the evaluation team should ask itself:

- Does the user understand the requirements?

- Will the construction team understand them as well?

- Are the requirements stable and under configuration control?

- Is it clear what process should be followed to resolve any conflicts in requirements?

- Is it clear who is authorized to adjudicate such conflicts and to change the requirements?

- Are any problems detected in the requirements clearly described and the facts communicated to project management?

- Are there any "sacred cows" lurking in the requirements? To whom does the sacred cow belong? To the customer? Marketer? Engineer? Management?

- Are the requirements robust? That is, will the system specified be adaptable to the likely (and inevitable) changes?

- Is it clear what the construction team must do in order for the project to succeed?

- Is it clear (or beginning to be clear) how the product should be integrated and tested?

In practice, the relative importance of any individual topic in these checklists will vary according to the system being constructed. The reader should also apply common sense when evaluating the requirements using these (or other) checklists. Checklists should be used as memory joggers to stimulate, not replace, thinking.

The requirements evaluation process should result in a report from the construction team (signed by the team leader) to the lead requirements developer, with a copy to project management. The size of the system being constructed, its complexity and criticality, and the volume of problems found should dictate the length of the report (which can be a one- or two-page memo).

In mission-critical applications, the evaluation process and report should be viewed as a *verification and validation* effort [Christensen and Thayer, 2002]. The evaluation report or memo constitutes the acceptance (or unacceptance) of the requirements by the construction team. The team and its leader should be comfortable with the requirements and their understanding of them. They should now have a clear vision of what the product is supposed to do and any constraints or other considerations that will drive or empact the construction activities. They should have detected and communicated any problems with the requirements.

Ideally, the architecture team should have done this during the transition from requirements to design. However, in practice, there is always a certain amount of defect "leakage" from one activity to another. After the lead requirements developer has had an opportunity to review and digest this report/memo, the three players should meet to adjudicate (under the leadership of project management) the concerns, and produce and assign any necessary action items.

The action items are the mechanisms that formalize any concerns and should only be used to document problems that really matter to the future of the product. Responsible individuals should be assigned to each action item and due (closure) dates established based on the time and effort needed to address the problem and the date that closure is required by the project schedule.

4. Design Evaluation

Having gained an understanding of the requirements, the construction team is now prepared to evaluate the architecture. While it could be argued that the requirements evaluation process should serve *primarily* as a learning exercise for the construction team, it is clear that evaluating to design is an organic part of the construction process. Namely, the construction team must review and understand the description of the system if it is to build it. In addition, since it is commonly the first independent group to examine the architecture, it is more likely to find problems with the design. Before beginning the evaluation, the team leader should ensure that team members:

- Understand the requirements.

- Understand the design representation, whether it is flow charts, data flows, functional flows, class hierarchies, or state-transition diagrams.

- Are sufficiently familiar with the application domain.

- Have some "scars" from earlier projects.

Depending on the size and nature of the product being developed, as well as the developmental process being followed and the methods and procedures used to construct the product, the design evaluated will be more or less tightly related to the construction methods. In some cases, the construction team will begin work immediately after the *architectural* description is completed. In many cases, intermediate levels of design must be derived from the architecture before the construction activities of detailed design, code and unit test, and integration and product test can effectively begin.

4.1. Evaluating the architectural description.

The *description* and its *elaboration* are the keys to the construction process. The architectural description is the first, most abstract representation of the system being constructed. It provides high-level views of how the system provides the facilities and capabilities needed to satisfy the requirements. It provides the conceptual framework into which the components will fit. It determines what components are built, how they are integrated to form the whole, and how the system will be tested during integration. IEEE Standard 1471-2000, *IEEE Recommended Practice for Architectural Description of Software-Intensive Systems*, defines the architecture of a software system as:

> *"The fundamental organization of a system embodied in its components, their relationships to each other and to the environment, and the principles guiding its design and evolution."*

It should be noted that the use (Annex B of the standard) of the word *component* in the quote does not necessarily imply the actual code and data components of the system. The usage in this context is more general, including the "...not physical, but instead, logical components." To clarify this, the annexes use the broader term *elements*.

The Standard then discusses 13 uses (numbered "a" through "m") of an architectural description, two of which are centrally concerned with the construction of the system. These uses are:

- "Input to subsequent system design and development activities."

- "Input to system generation and analysis tools."

There are a number of viewpoints and uses of the architectural description and correspondingly, a variety of representation methods. As a practical matter, two or three viewpoints are usually all that a construction team can work with when building the system. Annex C of Standard 1471-2000 lists three very common viewpoints:

(1) *Structural*

- What are the computational elements of a system and the organization of those elements?

- What software elements compose the system?

- What are their interfaces?

- How do they interconnect?

- What are the mechanisms for the interconnection?

(2) *Behavioral*

- What are the dynamic actions of, and within, a system?

- In what kinds of actions does the system produce and participate?

- How do those actions relate (ordering, synchronization and so on)?

- What are the behaviors of system components? How do they relate?

(3) *Physical interconnection*

- What are the physical communication interconnections and their layering among system components?

- What is the feasibility of construction, compliance with standards and evolution?

4.2 General design representation and evaluation.

There are varieties of methods for describing designs at all levels—from the architectural, to the intermediate, to the component. These include but are not limited to:

- Data flow designs, which demonstrate processing by focusing on the data needed by the functions and how they flow between the system components.

- Structure charts that show the initiation of a function starting at the top of the chart, with detailed subfunctions performed in lower levels of the chart.

- State machine representations that decompose functions into a nested series of state transitions between subordinate state machines.

- Object-oriented (OO) methods that focus on hierarchies of class objects composed of data and the methods used to access data and operate on them.

These approaches are not necessarily mutually exclusive. Thus, an OO method could be selected to represent and later implement a design whose internal mechanisms operate as a series of nested state machines, with the class and the inheritance properties of the design used to provide specific methods of computing states and performing transitions. Similarly, a dataflow architectural design structural description might be supplemented with a timing description and then elaborated with structure charts at the intermediate levels.

Independent of the design representation method, the design should be able to satisfy many criteria. Expanding upon the three viewpoints for viewing the architecture, six categories can be identified —

(1) ***Structural and behavioral*** — Does the designs look and feel right? Is it balanced? Is it justified?

(2) ***Feasibility and realism*** — Is it possible to build the design in the real world. Has the approach to do so been thought out?

(3) ***Functional completeness*** — Does the design satisfy all the stated functional requirements?

(4) ***Interoperability*** — Will the design operate in the stated environment without degrading its own operation or that of other systems?

(5) ***Suitability*** — Does the design satisfy the reliability, usability, security, and maintainability requirements?

(6) ***Readiness for implementation*** — Does the design contain the information needed for the construction team to effectively proceed? Is the architectural design at too "high a level" to allow the direct design of individual functions, procedures or subroutines? In other words, is an intermediate level of design needed before detailed component design can proceed?

When evaluating each of these six areas, many more topics than are detailed here should be examined. Many of these apply to all levels of design, although some are more appropriate to certain levels than to others.

(7) ***Structural and behavioral*** —

- Is the layout of the components and their interactions clearly specified?

- Are the class or data hierarchies appropriate? Do they relate to the application clearly?

- Are the class or data hierarchies sufficiently extensible? Are they too generalized?

- Is the internal protocol (procedural invocation or messaging) described adequately?

- Is the coupling of the elements appropriate and are the individual elements cohesive? (*See Section 4.4 for further discussion about these dual concepts.*)

- Is a justification for the design provided? Is it well thought out? Are trade-studies referenced? This is especially important at the architectural level.

- Is the level of detail consistent across the system?

- Does the level of detail require the interaction of too many (or too few) components or elements to accomplish the functions of the system?

- Are the interfaces to external systems and the user clearly visible and are designs appropriate? Again, this is especially important at the architectural level.

- Are major subsystems (such as user or database management interfaces) appropriately decoupled from the other parts of the application?

- If the application is multithreaded, how does the design handle deadlocks, stalls and race conditions?

- Are good information-hiding practices followed? Is data properly encapsulated using, for example, access routines or private methods?

- Can you "see" the application in the design or is it buried under a jumble of acronyms?

(8) *Feasibility and realism* —

- Are line-of-code budgets assigned to each component or element?

- Can the team construct the system within the allocated schedule?

- Are execution time and memory utilization budgets assigned to each component?

- Is the approach to operating within the specified execution environment clear and satisfactory?

- Is the need for, and approach to using, any application framework clear? This is especially important at the architectural level, although the approach to using a framework must be addressed at all levels of design.

- Have all components that will be bought, rather than built, been identified? Is the rationale valid? Is it consistent?

- Have all components that will be reused from other applications been identified? Is any necessary adaptation specified?

- Is it clear how the purchased or reused components will be qualified for use in the system and how they will be integrated into the whole? Are they included in the memory and timing estimates?

- Does the organization have the experience and talent needed to implement this design? Is training or hiring required?

(9) *Functional completeness* —

- Are the system goals clearly stated? Are the system-level objects (code and data) described? Are the names reasonable?

- Is the processing to be performed in each part of the design described adequately?

- Are the organization and contents of all databases documented?

- Are all the functions described in the requirements satisfied?

- Are memory and size code estimates provided for each component?

- Are all I/O described? Are they encapsulated by appropriate components, such as objects?

- Are the key algorithms of the system and components provided and their selection explained? Are trade studies/white papers provided or referenced?

- Can you "trace" or "walk through" the operation of each system function through the design? The level of detail of the trace should be consistent with the level of the design.

(10) *Interoperability* —

- Does the application conform to the operating system and I/O protocols of the execution environment?

- Will the application operate consistently across the range of specified environments?

- Will the operation conflict with other applications present in the execution environment?

- Will the operation be excessively vulnerable to the misuse of shared services by other applications?

- Is the use of all shared services consistent with the mandates of the provider of those services?

- Will the service over-occupy the CPU, screen, printer, or other system resource?

(11) *Suitability* —

- Does the function satisfy the reliability, usability, security, and maintainability requirements?

- Is the approach for using dynamic memory clear, along with how error conditions will be detected and dealt with?

- If dynamic strings are used, are sizing estimates provided, and is it clear what should be done if they are exceeded?

- Are errors in input detected at the point of entry into the system?

- Are error messages standardized, with a mechanism for automating their display?

- Is unnecessary coupling of system functions within the code avoided, so the functions can be modified independently?

- Are boundary conditions explicitly tested?

- Is the number of points of entry into the system minimized, so as to simplify the task of providing security?

- Are unnecessary features present, which can degrade reliability and security, and confound maintenance?

- Are user-dialogs natural from the viewpoint of the user, avoiding the entry of repetitive data across multiple screens?

(12) *Readiness for implementation* —

- Is the level of detail adequate, enabling individuals or small groups to effectively proceed with their individual component design, code and test assignments? If not, what method will be used to elaborate the design so this can be done? Does the organization process model allow for this intermediate step? Does the schedule allow for this intermediate step?

- Can a detailed implementation plan be developed from the design information?

- Can integration planning be started with the information provided?

Depending on the nature, intended use and size of the application being constructed, more or fewer of these questions are relevant. As with earlier checklists, they should be used to stimulate thought about the design and the construction activities to follow. As with the earlier evaluation, the lead constructor should generate a report or memo to the lead designer and project management, laying out any concerns.

After the lead designer has had a chance to review and digest this report or memo, the three players should meet to adjudicate (under the leadership of project management) the concerns, and produce and assign any necessary action items. The action items are the mechanisms that formalize concerns. As with any action items arising out of the requirements evaluation, responsible individuals should be assigned to each action item, due dates should be established based both on the needs of the construction schedule (see the following section) and the realism of actually closing the action item by that date.

4.3 Intermediate levels of design.

We have raised the possibility that the architectural description alone cannot provide enough detail to directly support component design and coding. This does not necessarily mean that the architecture design is incomplete or wrong. It is most commonly a reflection of the sheer size and complexity of the system. As a rule of thumb, systems larger than 10,000 lines of code require another intermediate level design.

Most applications contain approximately 10 major subsystems (for example, Word has nine command groups located across the tool bar). Commonly, each of these is in turn made up of roughly 10 functions that, for a simple application, could be accomplished in a single routine of code. If a typical routine is 100 lines of code long (nonblank, non-comment lines), then the system will be made up of $10 \times 10 \times 100 = 10,000$ lines of code.

In this case, the intermediate design is simply the breakdown of the 10 functions into sub functions, or of 10 Meta classes into specific classes for 100-based methods. If inheritance is properly used, the expansion rate (10:1) at each step should be reduced. In either case, how to perform the intermediate design will be obvious from the architecture and will require minimal effort to accomplish on the part of the construction team. It is now possible that the detailed design of a 10,000-line of code application can be derived directly from the architecture.

If, however, the functions or classes within the subsystems are more complex than this, they will themselves be broken down into an intermediate level of design, which will introduce another factor of 10, allowing the system to grow to 100,000 lines of code. In this case, an intermediate level of design is necessary.

In cases where an intermediate level of design is needed, the higher-level design entities would not be called components, but would be referred to by various names:

- Modules (especially in the Modula language)

- Packages (especially in the Pascal and Ada languages)

- Tasks (especially in real-time systems)

- Meta classes (in systems implemented in C++)

The methods used to design the internals of the modules, packages, or tasks, or to derive the subclasses, are as varied as those listed earlier. Again, hybrid methods are often used. For example, the architectural design for a real-time system might use data flows to describe the interactions of the tasks that make up the system, while the tasks themselves could be designed using structure charts. Alternatively, the tasks could be designed using an object-oriented methodology, whose message-based internal designs would be a natural match for the message-passing methods used to communicate between tasks in modern real-time systems.

Regardless of the method used to create and document the intermediate design, the goal is always the same: to produce detailed information that is used to design and build the actual software components (routines, data and objects) that make up the system.

However, care should be exercised when choosing the intermediate design representation to ensure that it is a good fit for the tool set, the application and the method used to represent the architecture. Care should also be taken to ensure that the intermediate design activities are represented in the schedule.

Finally, the intermediate design must be evaluated using a process similar to the one used to evaluate the architecture, but more specific to the level of detail. In addition, the names of the components need to be carefully thought out, so the intent of the design can be easily communicated. Appropriate names will also ease the integration and maintenance tasks. Criteria for naming components (routines and data) are provided in Section 6.1 of this paper explaining detailed design processes.

4.4 Cohesion and coupling as design criteria.

Cohesion and *coupling* are twin general concepts that should be used when evaluating the architecture, as well as in performing and evaluating any intermediate designs and ultimately, the code itself. They are useful when designing interactions of classical data and code components, or objects, the data they contain, and the methods used to operate the data. At a more abstract level, they can be used to evaluate relationships between classes, as well as the properties of the class internals.

The word "coupling" has been in continuous use for over 25 years, while the term "cohesion" was previously known as "strength." These concepts were introduced by [Myers, 1975] in his book, "Reliable Software through Composite Design." Both concepts are applied to an item (package, data type, routine, object, or methods) by comparing parts or properties of the item, which are simply referred to as "things" below.

The cohesion scale indicates how closely two items are related. The two items must come from within a single item (a package, a data type, an individual component, and a class). On a worst-to-best scale, cohesion takes on the following seven values [McConnell, 1993], [Myers, 1975]:

- *Coincidental cohesion* — There is no meaningful relationship between the two things (package, data type, object, or routine), like the proverbial kitchen sink. They just happen to be present in the same time. This will, if nothing else. makes it hard for integrators or maintainers to understand what is going to occurr in the process.

- *Logical cohesion* — Multiple parts are contained in the item but only one of them is used at any single point in the operation of the system based on a conditional (that is, a Boolean) flag. This usually results in tricky code, especially when changes are applied. It also requires that the flag is present somewhere in a module interface, in the data type, or (worst case) buried in the code.

- *Procedural cohesion* — Multiple parts, methods or data are contained in an item to ensure that a specific order of processing is maintained; however, the exact function of the item is unclear. The problem is that if the order changes, the item will need to be torn apart and recoded to change it. This tends to happen more with code than with data, where the items are more likely to be determined to be coincidental.

- *Temporal cohesion* — Multiple things are contained in an item solely because they are used or operated on at the same time. Myers called this "classical strength." In the case of packages and classes, this is not always a "bad thing" with initialization routines or destructor methods being classical examples. However, if done inappropriately, the package or class may later have to be broken apart to make changes. Depending on how data was partitioned, this can be a serious undertaking.

- *Communicational cohesion* — The parts of the item all influence one another. In the case of data, this means that changing one field of a record will result in changing another. For packages or routines, it means that the parts of the item influence one another through data items.

- *Sequential cohesion* — The item does not implement a complete function. It contains some things that must be performed in a series of steps in a specific order, and that share information (usually data) between the ordered steps. The fact that the item does not implement a complete function means it is *less likely* to create the maintenance problems that occur with items that have procedural cohesion. Myers called this "informational strength."

- *Functional cohesion* — All the parts (code or data) of the item are used to accomplish a single function with no side effects.

It should be appreciated that a package design or a class could have good cohesion (e.g., functional) but its internal components (procedures or methods) do not. Likewise, the data types used by the package (and available to its functions or procedures) could have good cohesion (e.g., communicational), but the internal data types used by the lower-level components might not. Of course, the worst thing that can happen is to have large entities (especially components or data structures) with poor cohesion.

Coupling is a dual concept to cohesion. While cohesion measures how two *things* within an item are related, coupling measures how two *items* depend upon each other. From worst to best, the types of coupling are [McConnell, 1993], [Myers, 1975]:

- *Pathological coupling* — One of the two items alters a property of another in an unnatural way. For example, one routine modifies a pointer used by another in a non-explicit manner. If the pointer is passed as a parameter (which would then be explicit), each modification would be okay. Objects could have this problem, since they are often implemented by applying pointers to their methods. Myers called this *content coupling*.

- *Global-data coupling* — The two items influence each other through, or are influenced by, data items that they can both explicitly access but which do not pass through an interface. The access does not have to be symmetrical. Myers called this *common coupling* and *external coupling*, to distinguish between the two situations.

- *Control coupling* — One item uses the other in a highly directive manner. For example, one routine calls another and passes in a flag that will cause a specific block of code to be skipped. This means that the two routines are *not* independent of each other.

- *Data-structure coupling* — Two items are coupled through a common data structure to which access is restricted so it is not global data. For example, a pointer to a data structure is passed from one routine to another, rather than passing, the data structure itself. The receiving routine does not have direct knowledge of the data structure location unless the parameter is passed to it. Myers referred to this as *stamp coupling* because the receiving routine is "stamped" with the information needed to access the data.

- **Simple-data coupling** — All data needed is explicitly passed by value, not by reference. This is the lowest (best) level of coupling. Myers called this *data coupling*.

It should be realized that building a system composed entirely of simple data coupled packages, each possessing a functional cohesion, with a particular programming language might well be impossible. It is sometimes necessary to comprise cohesion, for example, to reduce coupling.

The twin scales of cohesion and coupling allow us to evaluate systems, data items, packages, and individual components in a quantitative manner.

5. Detailed Construction Planning

So the evaluations and other preliminaries are done! Any intermediate levels of design have been completed. Finally, we can proceed with construction itself. However, a few messy details must be completed. First, the results of the evaluations need to be digested by project management, the requirements development team and the architectural design team. Second, they have to respond to any material issues, which should have been captured in the action items mentioned earlier. Generically, there are three possible outcomes for the evaluations:

(1) **Case I** — *Describes showstopper action items.* Only minor cleanups are needed to close the action items. The core requirements and the design will not change. The construction team can get to work, laying out due dates in the schedule for the individual action items. These are usually dictated by a combination of the integration schedule and the need for key personnel in other areas.

(2) **Case II** — *Describes showstopper action items that affect a few localized and distinct areas of the design.* Correction of these will not affect other areas. The construction team can get to work on the areas that are not impacted. The schedule plan should be segmented accordingly and a clear need date for resolution of the problems must be determined based on the integration and personnel assignment schedules. The plan should identify the need for a re-estimation of the cost and schedule budgets when the action items are closed.

(3) **Case III** — *Describes showstopper action items that affect most of the design, making it very risky to continue with any construction activities.* The construction activities should be suspended while the action items are being resolved.

The condition that occurs in a given project will affect the approach taken to laying out the detailed construction schedule, with the third condition forcing the suspension of virtually all construction activities. In the first and second cases, the detailed planning of the construction activities can proceed, with the caveats noted above. If it is believed that the second condition is the case, great care should be taken to ensure that, in fact, the problems are truly localized to a few areas of the design. There is a natural tendency to try to "get on with it," which often leads to self-delusion. This can lead to wasted effort, bad relations with the customer, and an erosion of employee morale.

5.1 Understanding the scope of the construction effort.

The plan for the construction effort is based on four criteria:

(1) The requirements that the product must satisfy. These should be documented in the requirements, the architectural design, and any intermediate design, as discussed above.

(2) Other requirements that the project must satisfy. These will usually be found in the contract with the customer or in a statement of work (SOW), but they may derive from strategic objectives belonging to the developing organization. Incremental deliveries are a common example of a project requirement that will influence both the structure and the duration of the construction schedule.

(3) The approach to integration and deployment of the products. Integration is discussed in Section 8.

(4) The available staff, methods (processes and procedures) and productivity that can realistically be expected during construction.

5.2 Contents of the construction plan.

The detailed construction plan should contain four elements:

(1) A detailed schedule showing:

- Who is assigned to perform the intermediate or detailed design of the component and when (date of starting the activity and completing it).

- Who will review the intermediate or detailed design and when.

- Who will prepare the component test plan, who will review and accept the plan and when.

- Who will conduct the component tests and when.

- When the component is needed to support the integration activities.

(2) A description of the methods and procedures that will be used to perform the technical work including:

- ***Detailed design and coding conventions with testing standards*** — Together these are usually called *programming standards*. The purpose of programming standards is to highlight exceptions so they can be judged on their merits; standards should not be absolute pro-hibitions; nor should exemptions be granted routinely.

- ***Procedures and rules for conducting detailed design, code, and test case reviews*** — The intent is to ensure that the work products listed above comply with the relevant standards.

- ***Any quality assurance (QA) policies and procedures that should be applied during the effort*** — This chapter will restrict itself to how peer reviews can help the individual engineer improve their work product. Other chapters of this book series discuss QA in the broader context.

- ***Procedures and rules to be followed when managing the configuration (versions) of the detailed designs, code and test cases of the components during construction, as well as during the transition to the integration effort*** — Other chapters of this book series discuss the configuration management in more detail. This chapter will restrict itself to a discussion of where to apply configuration control practices during construction.

(3) A staffing and labor expenditure plan showing:

- The number of hours planned to be expended on a periodic basis (daily or weekly, depending on the total duration of the project).

- Any key skills that are needed by the project but are not available when the construction plan is created, including when they are needed.

- A plan for how these skills will be acquired, through either reassignment of personnel, hiring or training.

(4) A listing of what resources (computers, office space, software tools) are required and when.

Not all projects require the same amount of planning. In some cases, resources are not an issue. Each developer need only have access to his or her desktop computer. As another example, the total scope of the project allocates two months for 10 people. In that case, if all the key skills are available, the staffing plan could simply be a graph showing the cumulative or weekly expenditure of labor. I hope that the organization has institutionalized code and detailed design standards that can be adopted with minimal customization. The general rule is to apply common sense, planning enough so you are comfortable with the plan and can explain it to others, including the members of your construction and management teams.

The one part of the plan that is less sensitive to the size of the project is the description of methods and procedures. The rule here is to include enough description so the construction team will know how to perform their technical tasks, while simultaneously satisfying any external requirements levied on the project in the areas (typically) of quality assurance and configuration management.

Of these four elements, the most difficult to develop is the detailed schedule. It should be developed by estimating the effort (expressed in hours of labor) and duration (expressed in hours from start to finish of a task as measured by a clock) required to perform each of the tasks necessary to construct a component. Using this data, which can be created in spreadsheet form, the network schedule showing the relationships between the tasks can be built using a network-scheduling tool.

5.3 Estimating the effort and duration of the construction tasks for each component.

The lead constructor should consider the following questions when preparing the detailed schedule and selecting the method to be used to estimate the time and effort required completing each task:

- What type of project is this? What are its deliverables and activities?

- Is this project type similar to previous projects?

- What kind of estimation techniques will be appropriate for this project?

- What are the risks that will affect the success of this project?

Historical cost and schedule data (sometimes called *past actuals* or a *project history database*) describe how much a project or previous projects actually cost and how much time was required to complete each deliverable. Ideally, this information should be broken down to the same level of detail as the current schedule. The lead constructor should consider the following questions when examining candidate projects, their products, development histories, and their historical cost and schedule data:

- Are the products similar?

- Are the technologies similar?

- Are the magnitudes of the project similar?

- Are the contract terms similar?

- Is the organization now more or less mature in its developmental practices?

- How much did this (the past) project cost to develop?

- How long did this project take to develop?

- How did the actual project cost compare to the budgeted cost?

- How did the actual project schedule compare to the estimated schedule?

- Was the project under budget or schedule?

- Was the project over budget or schedule?

- What areas experienced cost or schedule overruns?

- Why were costs or schedule overruns availible in those areas?

Realistically, the information needed to answer all of the above questions *exactly* is almost never available. Only by posing the questions to the lead constructor will insight into these issues be gained.

In practice, most software projects are estimated primarily by one method and then crosschecked, wherever possible, by others. The estimation is usually performed in an iterative manner, with an initial estimate based on the most generic characteristics of the project and subsequent estimates becoming more refined as the peculiar requirements of both the product and the project are included in the estimate. At the time of construction, this information must be available if the schedule is to have the desired validity and accuracy.

The estimation methods most commonly used are:

- *Parametric models*, which predict cost and schedule based on some parametric input, such as lines of code.

- Estimation using *documented past actuals*.

- *Rules of thumb*, which were often based on informal, undocumented, actuals.

For an in-depth discussion of these methods, the reader is referred to [Christensen and Thayer, 2002]. In the remainder of this chapter, for illustrative purposes, it will be assumed that the construction effort will achieve a productivity of three lines of code per hour of effort. Table 2 estimates productivity in terms of percentage of construction effort and effort hours per line of code for a construction team when performing each component of the major tasks of construction.

Notice that the total effort required to produce one line of code is 0.33 hours, which corresponds to a productivity of $1/0.33 = 3$ lines of code per hour. Ideally, the percentages of the total construction labor and the resulting hours of effort per line of code are based on histories of the organization. If this data is unavailable, the reader should consult [Boehm, 1981] or [COCOMO, 2000]. A table of task-by-task productivities should be used to develop the effort and schedule needed to construct each component of the design.

Table 2. Sample distribution of effort and hours required to perform construction tasks

Activity	Percent of total construction effort	Effort hours per line of code
Creation of detailed design	20	0.066
Review of detailed design	10	0.033
Post-review revision of detailed design	5	0.0165
Creation of component code	20	0.066
Review of component code	10	0.033
Post-review revision of component code	10	0.033
Creation of component test cases	10	0.033
Review of component test cases	5	0.0165
Execution of component test cases and rework of component to resolve any resulting anomalies	10	0.033
Construction total	100	0.330

To illustrate the use of Table 3, suppose that the product contains a component whose size is estimated to be 500 lines of code, the activities listed in Table 2 are all performed sequentially, the organizational processes require that detailed design and code reviews are attended by four individuals, and test case reviews are attended by three. The following would therefore be the effort and schedule required to complete the construction tasks.

From an effort standpoint, approximately 123 hours is spent creating and fixing the product, while roughly 42 hours of effort are spent finding the problems. From a schedule standpoint, however, the situation is somewhat different, with only 11 of 134.75 hours of the total lapsed schedule time being expended in the review process. The reason for this, of course, is that between three and four individuals are involved in the review processes.

If it is possible to assign the task of designing and coding the component to two individuals instead of only one, the effort would not be reduced, and in fact it would increase slightly due to the need for the two individuals to communicate. However, the schedule would probably decrease by approximately 33 hours (the sum of one half the detailed design and coding schedule spans).

Applying more than two individuals to code a component of only 500 lines of code is unlikely to be productive, due to the increased amount of communication required to coordinate their efforts technically. A more common occurrence is the "golden hands" phenomena, where a small number of developers have much higher productivities than the rest of the staff.

Table 3. Example of effort and schedule hours required to perform construction tasks for a 500 lines of code component

Activity	Effort hours	Personnel required	Schedule hours
Creation of detailed design	3.0	1	33.0
Review of detailed design	16.5	4	4.125
Post-review revision of detailed design	8.25	1	8.25
Creation of component code	3.0	1	33.0
Review of component code	6.5	4	4.125
Post-review revision of component code	6.5	1	16.5
Creation of component test cases	16.5	1	16.5
Review of component test cases	0.25	3	2.75
Execution of component test cases and rework of any resulting anomalies	16.5	1	16.5
Construction totals/averages	77 (Total)	1.9 (Avg)	131.75 (Total)

Such individuals can often complete tasks in much less time and with the expenditure of much less effort than the norm. Where to apply such talent is often not obvious without taking a holistic view of the schedule. This view is provided by a network schedule, in particular by the "critical path," which network schedules make highly visible.

5.4 Creating and using the schedule network.

Creating the network schedule requires the availability of three components:

- The time spans (the "schedule hours" column of Table 3) for each task of each component in the design and the corresponding availability of personnel.

- Identification of any external conditions or constraints, such as incremental deliveries *to* the customer, data deliveries *from* the customer (such as test data), or closure dates for action items.

- The dates by which the individual components or groups of components are needed by the integration effort.

It is important to differentiate between the *logic* of the schedule and the *finish dates*. Figure 1 shows a very simple network schedule.

The boxes represent tasks and their durations, which are respectively denoted by Tx.y and Dx.y. The arrows indicate that the preceding task (the box at the beginning of the arrow) must be completed before the following task (the box at the end of the arrow) can be started. The circles represent start dates and end dates for the entire network and are respectively denoted by dS and dE. The X represents an external event (denoted as Xy), which will occur at date dy.

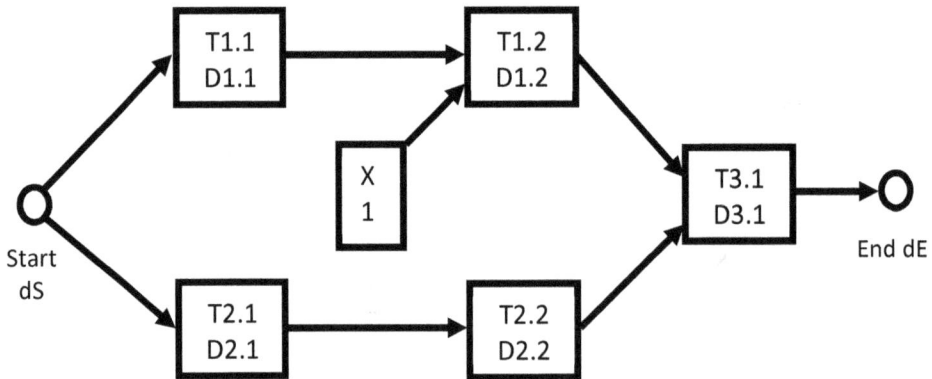

Figure 1: A simple network schedule

By developing the logic in this manner and "loading" it with the task durations and resource requirements, the detailed construction schedule can be created. In addition, if the network scheduling tool supports it, the resource types (skill grade of developers, for example) needed to perform each task should be specified and corresponding resource pools created. The scheduling tool will then attempt to apply the resources to the corresponding tasks in a manner that minimizes the total duration of the schedule.

Invariably, this is an iterative process for projects of any complexity. During the construction phase, however, there is a natural sequential structure to the tasks needed to build each component, as shown in Table 2. The only factors making it complex are the need for delivery of specific components to the integration effort by specific dates (to support product deliveries), any critical action items from the requirements and design efforts that must be resolved, and the availability of suitably skilled personnel.

Finally, the network should be updated on a regular basis to show the actual dates tasks are started and completed so the future status of the construction effort can be forecast and corrective action applied as problems arise. As effort is expended and tasks are completed, it will be possible to evaluate how well the project is proceeding by comparing the projected and actual task durations, effort expenditures and completion dates [Christensen and Thayer, 2002].

6. Detailed Component Design

With the component-level requirements in hand, the construction planning completed, and the individual developers assigned to their tasks, the design of the individual components can begin. This activity is well described by the term *internal design* [McConnell, 1993] to distinguish it from the other, earlier design activities. In addition to designing the internals of the components, each component must be assigned a name, unless this was done as part of an intermediate design activity.

6.1 Naming components.

The name of a component should be representative of what the component is and does. Thus, if the component is a function (that is, it returns a value), its name should reflect that fact. If it is a procedure (that is, does not return a value) its name should reflect that as well. For example, suppose a routine is a function that returns a reference (a pointer or an index) to the next record in a list satisfying a condition, starting from a given position in the list. A reasonable name for this would be:

nxt_matching_rec_ptr (current_ptr, match_val)

If the design (or the implementation language) required that this action be performed by a procedure a reasonable name would be:

get_nxt_matching_rec_ptr(nxt_ptr, current_ptr, match_val)

The distinction is the use of the verb "get" for the procedure. The use of a verb as the first part of the name emphasizes the action the procedure takes. In the case of the function, the first part of the name is an adjective, as part of a noun phrase. A simple noun would be better, if an appropriate one could be found. This is especially true for data items, whose names should almost always begin with nouns.

Rules to follow when selecting a name for a routine, data structure or object are:

- Primarily: Is the name representative of what the routine does in the application?

- Does it conflict with something else in the system or environment? Changing the case of letters is not an acceptable solution. Find a unique and representative name.

- Is it reasonably short (since it will be typed, retyped and modified)?

- Is it long enough to be representative and not be cryptic?

- After selecting a name, ask someone else what they think it suggests.

- Does it make sense in the context of the problem domain? Can you write a reasonable sentence using this name?

The project should provide naming guidelines (including names that should not be used) to the construction team. Abbreviation rules should be part of these guidelines. In the example above, "pointer" was abbreviated by "ptr," "next" by "nxt," and "value" by "val." It may seem trivial to save only a few characters, but character spaces add up, and characters should preferably be used elsewhere in the name where they are needed for continuity.

For more information explaining naming routines and data, the reader is referred to [Ledgard, 1987] and [McConnell, 1993].

6.2 Designing the data.

The data that flows between and within components is intimately connected with the actions performed by the components themselves. Niklaus Wirth phrased it concisely when he wrote his classic book titled *Algorithms + Data Structures = Programs* [Wirth, 1976], which offers an excellent discussion of the relationship between the actions a program are to perform and the things they are to operate on. The move toward OO design approaches and languages continue this trend.

Any system contains multiple levels of data. At the architectural level, the data types or classes will be very closely related to the requirements of the application and the operating environment. At the intermediate design levels, new data items or subclasses will correspondingly be needed to perform the processing or to implement the methods. Finally, at the internal level of an individual routine, data items will be needed to perform the actual processing, although this will rarely be of a new type.

There are a number of criteria that should be applied when designing data types, classes or objects for use within and between components:

- Will the new type or class have single or multiple instances? Will it become part of a super-type or class?

- Scope of access to the type or class and to instances.

- Size of individual items and assemblages of items (linked lists, arrays and so on). It may be necessary to write test code fragments to see what a particular compiler does.

- Relationship of the items (fields) within a type or class. Are the coupling and cohesion appropriate?

- Nature of processing to be performed on the type, or the methods that the class will support.

- How the language allows access to the type across routine interfaces.

- Is the data readily (in time) available to the components that use it the most?

Two key concepts of modern software engineering practices are information hiding and abstraction, which apply to both code and data. In the case of data, these two concepts refer to:

- **_Information hiding_** — Information hiding involves *localizing the data to only those design structures (packages, modules or individual routines) that need access to the data. And then allowing the structures access only to the degree absolutely required. Information hiding can also entail reading or writing the data (sometimes referred to as *use* and *set*) by means of access routines and controlling the visibility into the interfaces of those routines.

- The opposite of this is referred to as *global data*, whose use is recognized as being a bad idea. The organization of the source files (where data types are declared) and how they are linked (as a function of the language) can have a major influence on how well this goal is achieved. Data access control is often viewed as one of the most basic "defensive programming" techniques. OO languages provide excellent support for this if used properly.

- **_Abstraction_** — Use the language semantics (such as fields of records or structures) to express and access the data in a natural manner, instead of forcing the developers to know all the details of data storage. This allows the data to be expressed (in the system) in a manner that is much closer to the application domain. Languages like Java, C/C++, Ada, and Pascal allow the developer to express the data in a natural way, with the compiler handling details pertaining to how the type is implemented. The class hierarchies and inheritance mechanisms of OO languages expand this concept significantly back into the design process.

It must be borne in mind that striving for abstraction and information hiding does not mean the code should be obscure or tricky. It means that the design and code should allow enough visibility into what the system is doing to be built and maintained in a natural way.

The criteria laid down earlier for evaluating the coupling and cohesion should be applied to the fields within data types. In addition to Wirth's book, the reader may wish to consult [Abelson, et al., 1985].

6.3 Specify the procedural interface.

A component's name should be assigned simultaneously with the design of its interface. This would have been done for some components when the intermediate-level design was produced but, invariably, some iteration is required. In addition, the need for lower-level routines is recognized. They must be named and their interfaces specified. There are two types of items in any interface: inputs and outputs, with occasionally an item being both. Modern languages have mechanisms for distinguishing between these situations. Languages such as C++

and Ada carry this one-step further, by introducing "private" interfaces and methods.

The project programming standards should specify the maximum number of items allowed in an interface. Usually a number such as 5 or 7 is used, with a prohibition against collecting unrelated items into structures to get around the maximum rule.

The coupling between components can be evaluated by examining the interfaces of the components, together with their detailed internal processing descriptions.

The detailed design of an individual component should include a list of all:

- Inputs
- Outputs
- Global variables referenced, and yes, some may be unavoidable
- External I/O and communications the component performs

6.4 Design the internals.

The internal design of the component entails up to what the aforementioned discussions have been leading. The internal design should describe what the component is expected to do in response to a procedural invocation or message. The goal is to describe the processing of the component in sufficient detail so the code can be written (after the design is reviewed) without replicating the code. There are several methods among others available to represent the internal design:

- Structured English
- Program Design Language (PDL)
- Mathematical formulae
- State transition diagrams
- Pseudo-code

Of these methods, the most common is PDL. When designing a routine with PDL, the following guidelines will be helpful:

- Use the PDL as abbreviated English, with the goal of communicating what the code is to do unambiguously.

- Do not use programming language syntax on a one-for-one basis. If you do this, you are writing the code prematurely, before you have given it adequate thought.

- Write a sufficient level of detail allowing the code to be written directly from the PDL.

- Document any assumptions about the data being consumed by the component.

- Remember that the PDL will be reviewed, so others must be able to read and understand it.

- Lastly, if the PDL is as long as the code (in lines), then it is too detailed.

- Draft the PDL using a computer, paper, whiteboard, automobile windscreen, or whatever medium makes you comfortable. Draw pictures of data structures and rework the PDL until you understand the problem and the solution.

Done properly, the PDL can be used to comment the code later. Some organizations prefer to capture the PDL in a separate set of files, while others prefer to use the PDL as part of a header for the component.

Components can be broadly categorized according to the primary nature of their processing. There are generically three categories, although most algorithms contain some aspects of all three:

- Logical algorithms

- Data access or processing algorithms

- Numerical algorithms

Thus, communication protocols are a hybrid of logical control and data processing, while GPS geo-location systems require precise numerical processing. Graphical systems primarily combine data and numerical processing. The literature of software engineering and computer science is rich in reference materials in these areas including [Wirth, 1979], [Knuth, 1981], [Press, et al., 1990], [Oppenheim and Schafer, 1989], [Pyster, 1980], [Kruse et al., 1991], [Lynch, 1985], and [Grimaldi, 2000].

6.5 Document the routine's design.

Design is never a straightforward, linear process. Invariably, as you work on a problem you learn more about it, gain insights into its structure and come up with better solutions. As the internal design of components progresses, insight is gained into how data should be structured [Wirth, 1976], and vice versa. This is natural, but at some point, you have to stop iterating and write the routine design so it can be reviewed by others.

The project programming standards should specify the format for the routine design. An example would be:

- Project name
- Routine name

- Reference to component's requirements

- Modification history (if not provided by an SCM system)

- Routine purpose

- Input parameters

- Output parameters

- Global accessed

- Constants used

- Hardware, timing or operating system dependencies

- Assumptions

- Description of internal data

- Description of processing in PDL

- Date

- Author

The resulting page or two of text should then be inserted into the project configuration control system (CM) and the review scheduled. The reason for inserting it into the CM system before the review is to enable the review team to know where to find it, and to have confidence that it will not change between design and the time the review is scheduled.

In addition to the formal configuration managed copy of the "final" (from the immediate standpoint of the individual developer) design, many organizations and developers maintain working level copies in a *unit development folder* (UDF). In earlier periods, the UDF was literally that: a manila folder containing working notes, draft designs, applicable memoranda, and so on. While much of this material can be stored online today, there is still a place for a hardcopy UDF. Other engineering and scientific disciplines use laboratory or engineering notebooks for this same purpose.

6.6 Review the design.

Other chapters of this book discuss the general topic of reviews. In the context of the detailed design, the review should be conducted by the developer's peers. A typical team of four to six individuals is used to review the detailed design of a small collection of components over a period of one to two hours. This usually limits the scope of such a review to components totaling at most 500 lines of code.

When conducting the peer review, it is useful to keep in mind that the review's purpose is to examine the design dispassionately and objectively. Useful questions to ask when conducting a review include:

- Are all the inputs and outputs specified correctly? Are they all used in the processing description?

- Does the description flow smoothly?

- Is the component turning out to be bigger and more complex than expected? Perhaps it needs to be further decomposed.

- Is it clear that this component solves the stated problem, and how does it do so?

- Is it clear that this component design solves *only* the stated problem?

- What is the component's cohesion? Is this the best it can be, given the requirements?

- How is the component coupled to the rest of the system? Is this the best it can be as well?

- If the component is a function, should it be? Likewise, if it is a procedure, should it be?

- Is the component's name descriptive? Does it satisfy the criteria given earlier?

- Is the component's internal data necessary and appropriate? Does it also satisfy the criteria given above?

- Does the component protect itself from bad (incorrect or with errors) input data? If not, where else in the system is this done and do you believe it?

- Does the component protect itself from internally generated bad data?

- Does the component use pointers? If so, how does it validate them?

- Does the component use dynamic memory? If so, where is it allocated and where is it released?

- Will the code, when written, be easy to change?

- Will the code, when written, be easy to test? Does the design include provisions for debugging "hooks" that match the capability of your test environment?

- Does the component implement the error processing approach of the greater design?

- If the component uses other components that return error codes, are the error codes used and are the errors handled? Are they ignored?

If the component implements mathematical formulae, are the calculations decomposed so as not to lose numerical accuracy in the intermediate results? Have the equations been checked to ensure that they perform as expected over their entire range? If this was done using a spreadsheet or other mathematical tool, do the numeric types of the tool have the same ranges and precision as those of the implementation language and computer system? The programming standards of the project should provide a checklist for the review team's use, together with a review report form so the results of the peer review can be documented. The minutes of the review should include a list of action items, documenting any issues to which the component designer must respond before the generation of actual code can begin.

The degree of emphasis given to some of the questions will vary, depending on the nature of the component and the programming language. For example, subtle problems with pointers and dynamic memory are very common in the C language, much more so than in Ada. Indirection and memory allocation are intrinsically difficult subjects. The book written by Daconta [1993] offers practical advice to C programmers.

7. Code and Unit Test

With the reviewed and approved design in hand, coding and testing of the individual components can begin. This final step is no more automatic than the earlier ones. The internal design produced in the prior step must be elaborated upon and mechanized. Problems undetected by the developer or by the review team will be discovered and must be solved.

In addition to writing the code line-by-line, the code must be visually organized (or laid out, to use the typography term) so the developer, reviewers and maintainers can readily orient themselves to the body of the component. Comments must be added to aid the developer, reviewers and maintainers, but not so many as to overwhelm and hide or obscure the code itself.

The major issues that arise in writing code that impacts quality during testing, deployment and maintenance are:

- *Satisfaction of requirements* — Does the component do what it is supposed to do? Does it implement the required methods or functions?

- *Flow of control* — In the early days of computing, this dominated all other factors. It is still very important.

- *Complexity* — Is the component more complex than the requirements demand.

- *Data* — Is the data being used according to its design or is the code coercing it to take on a value that is more natural for a different type?

- *Coupling* — Does the component have appropriate internal coupling?

- *Cohesion* — Does the component have appropriate internal cohesion?

- *Localization* — Are related operations close together in the component? Or, must the reader jump back and forth across pages?

- *Transparency* — Is it obvious what the code is doing or does it rely on "side-effects" to handle special cases?

- *Comments* — Are comments appropriate? They should not replicate the code. Nor should they be irrelevant.

- *Layout* — Can the readers rapidly orient themselves throughout the code to find things?

- *Length* — Is the component too long? Does the project include a standard for this?

- *Run-time errors* — Once the code is working as intended, does it detect and handle error conditions? It is one thing to indicate this in the design but another thing when indicated in the code.

- *Testability* — Will the code be easy to test with a reasonable number of test cases or does its internal design complexity require an excessive number?

Some of these questions are environmentally dependent. For example, some development environments support automatic test case generation and regression testing, whereas others do not. The testability question would have a different answer depending on the environment.

The question of how the code is laid out and commented is an important one because, while an automatic compiler will not care about visual organization, the humans who must read the code to review it and maintain it certainly will care. The project should include commenting and layout guidelines in the programming standards. Some general suggestions follow:

(1) Comments should:

- Not replicate the code.

- Indicate what the code is trying to do, that is, the intent of the code.

- Not be interspersed and interwoven with the code too densely; doing so makes it hard to find the code and follow its logic.

- Be simple and helpful.

(2) When laying out the code the developer should use:

- Blank lines to visually divide the code so readers can find things easily, much like paragraphing in normal writing. This may also indicate new components that should be separated.

- Indentation to highlight control structures.

- Blank spaces to highlight terms in expressions. Don't let the characters run together. Eyestrain will result when trying to read them.

- Format consistently. Don't keep the reader guessing as to what your style is.

- Place declarations at the beginning of the component. Don't declare items in the middle of the component.

Both [Ledgard, 1987] and [McConnell, 1993] discuss these topics in detail, with examples.

The matter of coding length is often a controversial topic. As a rule, it is good to keep the length to less than 100 lines, although there are always exceptions (such as simple case statements with lots of selector values, each of whose corresponding code is simple). Keeping the components small makes it easier to navigate inside them. If the reader has to keep flipping back and forth between pages or screens, it very easy to miss instructions or data and become disoriented. The author once wrote a moderately sized application without access to a printer. As a result, all functions and procedures were sized to take up one screen on the monitor.

It should be noted, lastly, that some of the items in the checklist (such as those for coupling and cohesion) are similar to questions asked about the components' detailed design. However, the viewpoint is different, as is the level of detail. Likewise, the answers to some of the questions have a strong influence on other questions. Thus, flow of control will influence complexity, testability, coupling, cohesion, and localization.

7.1 Flow of control.

The profession [Dijkstra, 1968] has recognized for some time that the best component designs are those that start their execution at the beginning and stop at the end, with a minimal amount of "jumping" around in between. It has been shown that all components can be written with three control structures to guide the processing from the beginning to the end:

- *Sequence* — That is, one statement follows another with no interruption.

- *Conditional* — That is, one block of code is selected from among a finite number of alternatives. If and case (switch) statements are examples.

- *Looping* — That is, a block of code is executed repeatedly until some condition is met. For, do-while, and while-do loops are typical examples.

When implementing the flow of control in a component, the following guidelines should be borne in mind.

(1) *For sequential code:*

- Is the sequence correct and natural?

- Does it read naturally from top to bottom?

- Are related lines adjacent to one another?

- Does the code make it clear what the dependencies are between the lines and the data items referenced by them?

Can the code be broken into multiple sequential blocks, each of which does its own independent processing function? If so, the component probably had temporal or procedural cohesion and should probably be broken into multiple components, so each one will have better cohesion.

(2) *For conditional code*:

- How complex is the logic? Can it be simplified? Look for alternative ways of expressing the logic, perhaps using a truth table to reduce the intrinsic complexity [Grimaldi, 2000].

- For any 'if-then' constructs, is it clear why the 'if' is present? Will the code within the 'if' block have any side effects on other parts of the component? Will the code execute correctly for both conditions? If there is an 'if-then', should there be a corresponding else?

- For 'if-then-else' constructs, is there any common code within the two blocks? Can code be moved out of the constructs so it only appears once?

- If there are any nested 'if-then-else' constructs, they should be examined with great care and even suspicion. Some languages (such as C and Pascal) handle the 'else' differently, depending on how the if-then' is terminated. Should a case statement be used instead, or the nesting divided between modules so the code is not as cluttered?

- Are there many sequential 'ifs'? This may indicate that a case or switch statement is more appropriate.

- For all of the above, is the condition being tested complex, involving several variables? If so, it is likely to be wrong, or at least hard to test and maintain. Perhaps it can be simplified logically [Grimaldi, 2000], or perhaps you should write a separate function that does nothing more than evaluate the condition?

- For case statements, are the cases ordered in a natural way? Should there be a default case? Is each case properly structured, simple and appropriately terminated? Does the need for a case statement arise naturally from the problem being solved?

(3) *For looping constructs:*

- Which type of loop should be used? For loops are naturals when traversing simple lists and arrays with simple exit conditions. 'While-do' and 'do-while' loops are better for conditions that are more complex. 'While-do' loops may never execute, and 'do-while' loops are guaranteed to execute at least once. Does this matter in your component?

- Is the termination condition well understood, clean and natural to the application?

- Does the loop perform only one function? Does it have side effects?

- Are there safety counters or flags used to avoid either infinite loops or illegal memory access?

- In the case of 'for' loops, is the name of the loop variable simple and appropriate?

- For 'while-do' and 'do-while' loops, is the exit (or continuation) condition simple? If not, apply the methods recommended for dealing with complex conditionals.

- Does the code within the loop have strong cohesion?

- Is any initialization code located immediately before the do loop?

- Is any clean-up code located immediately after the loop?

Exercising care when designing the flow of control within and between components will result in benefits, in the initial creation of the component, as well as during unit testing, integration and maintenance. There are, of course, other control structures in some languages. These include recursion, multiple exits (returns) from a routine and go tos. Recursion occurs when a routine calls itself. Sometimes it is the simplest way to solve a problem but can lead to subtle infinite loops or memory (runtime stack) overflow conditions. Any problem that can be solved by recursion can be solved by iteration. So there is never an absolute need for it.

Returns from multiple locales are allowed in some languages (C in particular). They violate the fundamental tenets of structured programming and are never absolutely necessary. The use of multiple returns is usually a danger sign, making maintenance and test very complex. Likewise, the appearance of a "go to" in a program written in modern language is usually a sign that you have programmed yourself into a logical box. Go back and rethink the logic.

7.2 Design the unit test cases.

After the code has been written, reviewed and compiled, it will need to be tested, either by the developer or by a separate test organization. There are many approaches to testing [Kit, 1995], but most organizations require that the individual developers test their own code. If that is the case, then it is reasonable to have the developer design the test cases for the component they are designing and coding. The term "unit test" is used here to avoid confusion with the way the term is used in the field. It has historically been called unit test because an individual unit of code (which we have called a component or routine) is being tested.

The general subject of testing is discussed elsewhere in this book series. This section will restrict itself to those aspects of testing that are most appropriate to unit testing conducted by the developer of the component. A good general introduction can also be found in [McConnell, 1993] and [Hetzel, 1988].

It is always beneficial to remember why an activity is being performed. In the case of unit testing, it is usually being performed to show that:

- The component (unit) performs the functions assigned to it by the design.

- Avoidable latent defects have been removed.

- The component is ready to be passed on to the integration team.

To accomplish these objectives the unit tests are usually designed to demonstrate that:

- Each function assigned to the component is indeed performed (and correctly) for nominal values of the inputs to the component.

- The internal code of the component is completely exercised. This means that all branches of control structures are taken but not necessarily in all combinations. However, all conditions that cause each control branch to occur should be tested, including each alternative logical condition.

- The component interacts correctly with any external entities, including called routines, code or operating system or hardware interfaces.

- Error cases are detected and handled correctly. In principal, this should be addressed by the branch coverage requirement but special attention is warranted for errors. After all, the condition was considered important and dangerous enough to be called an error.

- The code is tested with its external and internal data in the various states it can use.

How many test cases would be required to do this? To satisfy the first requirement, if the component performs N functions (ideally, this is one based on cohesion) then there should be N nominal test cases designed, each of which exercises one function in the component. Further, if the component contains an if statement:

if (*condition*) then

then at least two cases will be required, one to force *condition* to be true and another to handle the false. However, one of the two cases will already be covered in the nominal function test set.

If *condition* is simple (such as x < 10.0) then designing this new test case requires that the component inputs cause x to assume a value less than 10.0 or the complement, greater than or equal to 10.0. Which of the two is selected will be determined by which case is already included in the nominal functional cases. Thus, when the conditional is simple, two cases are required but only one is uniquely designed. However, if *condition* is compound, such as:

(x < 10.0) and (w > 12.0)

At least one more case is required, to test the *and* statement, making a total of three. If the operands of the *and* statement themselves contained logical operations such as *or*, then at least one test case would be required for each of those, and so on. Of course, a case is required to test the logical operators regardless of where they appear in the code.

Finally, if the code contains a *case* or *switch* construct, there should be one test case for each case, including the *otherwise* or *default* clause, even if none is present.

This level of testing is referred to as basis testing by McConnell [McConnell, 1993]. It is probably the minimal level of testing that should be performed at the unit level. The next level of testing would be pair wise condition testing [Phadke, 1989], in which each combination of pairs of conditions are tested. To illustrate, consider the compound conditional:

[(*condition1*) and (*condition2*)) or (*condition3*)]

Basis testing requires two unique conditions (in addition to the one resulting from the nominal functional test set). There are in fact eight possible conditions present when all combinations are considered, which requires seven additional tests instead of three. It is easy to see that this can rapidly get out of hand, as testing grows exponentially by the number of conditions. An alternative is to test the conditions pair wise, which requires a total of four test cases, one of which is the nominal functional case. Table 4 lists these four conditions.

In general, the pair wise conditional approach will result in linear growth in the number of test cases with better coverage than the basis method and only slightly more cost (three additional cases instead of two). Of course, if the routine will be used in a critical application, it would be appropriate to test all the combinations, depending on the penalty for incorrect operation. Ultimately, this argues in favor of simplicity: if the component is simple, it will be easier to test. If it is complex it is harder to write, harder to review, harder to test, and more likely to contain latent defects.

Table 4: Pair wise testing of conditionals

Test case	Condition1	Condition2	Condition3
1	TRUE	TRUE	TRUE
2	TRUE	FALSE	FALSE
3	FALSE	TRUE	FALSE
4	FALSE	FALSE	TRUE

To summarize, by designing one nominal test case for each function the component performs, and then by adding a test case for each of the control statements and logical conditions, a minimal set of test cases can be defined. Next, evaluate these to see if they force special events, such as external interactions and error conditions and add specific tests to cause these to occur.

In addition to code coverage, additional test cases should be designed that place the data used by the component into the states it can assume [Beizer, 1983], [McConnell, 1993]. This approach defines the following states that data can assume within the scope of a component:

- **Defined** — The item was written to but has not yet been read. This is sometimes known as *setting* the variable.

- **Used** — The item has been read.

- **Killed** — The item was written to (defined) but the value is now undefined. This could happen in multithread systems with shared (global) variables or it could simply be a for-loop counter, whose value is undefined after the loop is exited.

Earlier control coverage testing will ensure that all instances of a variable being defined (or set), used and killed will be tested individually. The next level of testing requires that each pair wise temporal combination of defined-used, used-defined, defined-killed, killed-defined, used-killed, and killed-used be tested. Of these, only two, defined-used and defined-killed, should really be required, as the others are either benign (killed-defined, used-killed) or generate obvious error conditions (killed-used, used-defined), which should be discovered in either the code review or during the coverage testing discussed earlier.

It is important to recognize the difference between planning the test cases (recognizing the need for them) and designing them (selecting conditions that will cause them to occur). Planning of test cases, especially when logical conditionals are present, is usually easier than designing them. Designing such cases entails inverse thinking, which is more difficult. For example, if a numerical value is internally computed by a complex algorithm and is then used to control the flow inside a component, it may be difficult to design a test case that causes both conditions to occur. This, of course, argues for simplicity.

7.3 Review the code and test cases.

After the code is written, the test cases designed (or at least planned) and the results placed into the project CM system, it is time to conduct the code review. Like the detailed design review, this is a peer review conducted by a small group of individuals who are well prepared in advance for the review. In many organizations, the code and test cases are reviewed separately, although this writer would argue that the code and test plan (which identifies what cases are needed) should be reviewed together, if for no other reason than to remind the group that the code will in fact be executed. Criteria to apply when examining code were given earlier in Section 7.

Some specific guidelines for the conduct of the review include:

- Has everything been drawn from the project CM system?

- Has everyone in the review prepared adequately for it?

- Has appropriate background material been provided, such as the detailed design of the component, or the definitions of any data structures accessed by the component?

- Does the code comply with the design?

- Do the review team members have appropriate technical backgrounds in coding, testing and design?

- Is there a mark-up copy of the code? This should be used to capture specific references to the code.

- Just as with a design review, there should be specific action items documenting problems seen with the code.

- Take it seriously and don't nit-pick matters of personal style. Focus on real issues.

- Finally, the action items should be captured and closed before the start of the unit test.

- The project should decide, based on the severity of the action items, if a new code review is required or if the developer can proceed directly to unit testing once the action items are closed.

It is always possible that the review team will realize that there is a problem with the internal design of the component, with the intermediate design of the system, or even a problem with the architecture. Should any of these problems occur, the problem should be raised with the appropriate parties and should be fixed.

Lastly, controversy always exists concerning how much compilation and testing should be performed before the code review. Some organizations require that the code NOT be complied before meeting for the review [Dyer, 1992], while others require that it be unit tested. The general practice is to compile the code before the meeting but do nothing else, based on the assumption that this eliminates numerous language-specific issues. Testing the code prior to the review can create many issues and defeats the purpose of reviewing the test plan and cases.

With the action items closed and the code revised and approved, either by a new review or by the project's technical leadership, it is time to test. Almost invariably, the unit test process uncovers problems not seen during the review, although usually detecting fewer problems than would be the case had there not been a review.

Because of the problems, changes will be made. Based on the complexity of the changes, the code will need to be re-reviewed at some level of detail. Errors of omission usually require little review, while those that require logic changes to the code often require review that is more extensive. The process of finding and fixing problems should continue in a controlled manner until the defects are eliminated. Components that make too many (more than two or three) trips around this loop should be viewed with alarm and pulled back for review at the design level.

8. Integration

The integration team is the "customer" of the unit-tested components. The project's approach to integration is important because it, in conjunction to the project's release schedule (to field testing or other deployment), determines the order (need dates) in which individual components must be completed. In other words, the integration plan generates a *pull* condition on the component construction schedule.

There are three basic strategies for integrating components and systems:

- *Top down* — The top-level components are completed first and tested using "stub" or "dummy" routines to simulate the lower level routines they will invoke.

- *Bottom up* — The bottom-level components are completed first and tested using "drivers" to simulate the higher-level routines that invoke them.

- *Flow* — A flow or thread of execution is traced through the system and the components that must be completed first. This can be very useful in multithread real-time systems. The actual thread could be integrated using a top-down or bottom-up approach.

Table 5 Christensen and Thayer [2002] lists some of the issues associated with the first three basic integration approaches.

Table 5: Pros and cons of integration testing
[Christensen and Thayer, 2002]

Top-down testing	
Advantages	**Disadvantages**
(1) More productive if major flaws occur toward the top of the program.	(1) Stub modules must be produced.
(2) Representation of test cases is easier.	(2) Stub modules are often complicated.
(3) Earlier skeletal programs allow demonstration (and improve morale).	(3) The representation of test cases in stubs can be difficult.
	(4) Observation of test output is more difficult.
Bottom-up testing	
(1) More productive if major flaws occur toward the bottom of the program.	(1) Driver modules must be produced.
(2) Test conditions are easier to create.	(2) The program as an entity does not exist until the last module is added.
(3) Observation of test results is easier.	
Flow testing	
(1) More productive if major flaws occur along a major flow of control or data. Best at finding flaws in architecture.	(1) Flows and impacted modules may be difficult to separate or recognize.
(2) Test conditions directly related to system operations/input requirements, hence easily understood and created.	(2) Modules from multiple groups may be needed to complete any individual flow, increasing planning burden.
(3) Test results can be easily interpreted and related to system output requirements.	(3) Partial versions of intervening modules are needed.

In some cases, groups of execution threads are required to implement a single feature, in which case the grouping creates a high level of integration based on the flow integration. In practice, many projects use a hybrid approach. This can be especially useful for systems with both low-level hardware interactions and high-level user interfaces to integrate from the bottom up (to test the interactions with the hardware) and the top down (to test the user interface), meeting in the middle.

Whichever approach is selected, the project should insure that:

- ***Integration is planned*** — Test cases should be designed and scheduled.

- ***Integration is managed*** — The progress of integration is tracked. Necessary resources are provided. Problem reports and the project CM system should be used to manage the configuration of the components as they are being integrated.

- ***The developers are involved*** — Unbelievably, some organizations try to integrate without involving code developers, while some developers resist supporting the integration.

In addition to these general issues, the integration test environment, the test cases selected, and the test data required to perform those tests are all critical. Obtaining representative test data sets and execution environments (especially for Net-based server applications and embedded control systems) can be challenging.

Performance problems often arise during the testing and integration process. That is, problems will arise if the testing is well planned and the test data and environment is representative. If they are not, then problems will arise in usage. Assuming the former condition (and trying to do all we can to avoid the latter), the classical areas where performance suffers are:

9. Tuning for Performance

The process of testing, evaluating and modifying the system to fix problems in these areas is referred to as performance optimization, or simply as *tuning*. It is a verity of software engineering that the first two are often coupled, with tradeoffs between them being possible and necessary. Most discussions of this topic focus on the first two aspects of performance [McConnell, 1993], [Jain, 1991], [Nemzow, 1994]. For some systems, all three are important and coupled, with accuracy improvements usually, but not always, requiring more memory and time [Kulisch et al., 1995].

All tuning should be approached with due care, bearing in mind the following:

- Is it necessary to tune at all?

- Do you really understand where the problems are and what causes them? The causes are referred to as *bottlenecks*.

- Is more than one kind of performance problem occurring, whether simultaneous or not?

- Do multiple situations cause the problem to occur?

- Is there anything you can do about it?

Finally, it is beneficial to remember that many applications are network or Internet based. For such systems, the impact of network loading and overhead must be considered as well [Nemzow, 1994a]. The application should have been designed to deal with these factors but the reality of the operational environment is often different from the specified environment. Or, it differs depending upon how the designers and the constructors interpreted it. Finally, it is possible that the application is a client of a server application and that the server may slow down or fail completely. In that case, the performance issue may be totally outside the client's control but should respond robustly to the condition.

9.1 Memory tuning.

There are a number of basic precepts that should be followed when designing or redesigning data structures and their relationships to code.

- Use an efficient representation. There are usually two pieces included in a representation: efficient use of memory and efficiency for the code to read and write to memory. Make sure you understand how your tool set will implement (allocate data items to the levels of bits and bytes) your data structures.

- Use only necessary memory. This may mean using dynamic memory. It may mean just using the simplest data type necessary, such as a character instead of an integer, or an integer instead of a floating-point number.

- If you are using dynamic memory, be sure that the deallocation function really works. Software history is full of examples in which the deallocation mechanism either did nothing or did not completely release the memory blocks it should have.

- If complicated logic can be replaced by a static table of results (a so-called "look up table") that takes up less space, then replace it. Doing so will also be faster and more reliable.

- When sizing static memory items, especially arrays, make sure you understand the usage patterns so you do not exchange a memory access problem for a data overrun.

- Are multitasking and multithreaded operations necessary? The whole community suffered with non-reentry of DOS long enough. Don't be the next one to make a similar mistake.

Again, memory use problems can often result in time (speed) problems. The reason for this is simply that neutrons and protons are nearly 2,000 times more massive than electrons. Disk drives (whether located on the machine executing the code or remotely sited) must move a lot of neutrons and protons to access a bit of data that would only require a much smaller number of electrons in the main memory. The solution is to keep the data that is frequently accessed in memory.

There are five ways to do this:

- Modern processors implement memory caching. Memory caching stores frequently accessed code and data in specialized high-speed memory. The problem is that most environments make it nearly impossible to explicitly force the processor to cache a specific block of memory. You may receive benefits from the processor cache, but you will probably not be able to do much to control the situation. (*See [Abrash, 1994] for a discussion explaining how to turn on and use caching for the Intel family of processing chips in Assembly language.*)

- The disk I/O of many operating systems uses caches and read-ahead disk blocks. You may be able to influence the block size that is used. Then again, you may not, especially if the data is not located in (logically) contiguous disk blocks. In addition, if the data is located on a remote server, you probably will not be able to influence the block size at all.

- If none of the above work, you can do local, application-dependent caching, retaining the data that is most frequently used by your application. This is more work than the other approaches, but it is guaranteed to yield the maximum benefit, provided it is really needed. This strategy works with both local and remote data. However, if the data comes from a shared database, then care should be taken not to "hog" the data by locking it for excessive periods. The author once saw runtime of the parts order subsystem of an ERP (Enterprise Resource Planning) system from more than 24 hours to 15 minutes using this technique.

- In extreme cases, you may need to force the data to reside on specific tracks and sectors of the disk drive. Real-time data acquisition systems sometimes must do this. Implementing this approach requires very low-level control of the device; this is usually done most safely with a dedicated device. As an alternative, the disk drive can be defragmented when the application is installed and a new file allocated with the necessary size. This should force the data onto logically contiguous track sectors. If the device driver was well written, tracks and sectors will be organized for efficient access.

- Overlap processing with I/O. This is often implemented by "double buffering," namely, processing a buffer already in memory while requesting the next buffer from the external device. While the external device is fetching the next buffer of data (which will probably take milliseconds), the application can be processing the current buffer. This assumes, of course, that the I/O subsystem can truly operate in parallel with the application code.

9.2 Algorithm tuning.

The duel to designing and accessing data is the code that consumes, creates or modifies it. In the previous examples, all the changes would, of course, require changing the code. However, the changes were not at the algorithmic level. Neither the design nor the code of the algorithm was fundamentally changed. Algorithmic tuning either requires a more efficient implementation of the algorithm, or its complete replacement with a new algorithm.

Typical strategies for tuning algorithmic code include [Abrash, 1994], [Bentley, 1986], [Bentley, 1988], [McConnell, 1993].

- If an expression or part of an expression occurs several times in a routine and its arguments do not change, consider computing the value once and storing it for use in subsequent code. This is referred to as the elimination of a common subexpression. It gives benefits only if it is inside of a looping structure that is executed many times.

- In a similar vein, simplify expressions. This will also aid maintenance.

- If a loop is executed a relatively small number of times, some time savings may be achieved by eliminating the loop and replacing it with sequential statements that do the same thing. This is referred to as "unrolling the loop." Doing so eliminates the overhead of incrementing and comparing the loop counter. However, most compilers implement this pretty efficiently today and most processors have instructions designed to make loops efficient. So, this rarely produces significant savings.

- For loops that are repeated, many times the work performed within the loop should be minimized. This can be especially fruitful if the code within the loop itself implements an algorithm that also involves looping. The author once improved the runtime performance of a system with a hard deadline from 300 percent of that required to less than 70 percent by redesigning the internal loop of a single routine that was itself repeatedly invoked from within a loop. The total change was only a few lines of code in the routine and one line of code in the loop itself. The reduction in runtime obtained from making changes of this type can increase significantly with the size of the data set.

- Order the case selectors in case statements according to their frequency of occurrence. Try to use contiguous values for selectors. Good compilers will produce a jump table..

- Put the most likely case of an *if* statement in the *then* block, placing the least likely in the *else* block. This may or may not offer paybacks.

- Research the literature about algorithms. Most problems are not unique, so there should be information out there that will help you or, at the very least, stimulate your creativity.

- Finally, consider converting critical sections to assembly language. Before doing so you should examine the output assembler listing produced by the compiler. A good compiler should do an accurate job, provided the algorithm is good to start with; so know your tools.

In all cases, the best instruments available for tuning are your eyes and your brain. Use both to guide testing before making changes. Do not use a shotgun approach to testing to find the problem. Design a logical series of tests to isolate the problem.

9.3 Numeric accuracy tuning.

Problems with numerical accuracy can sometimes be very subtle and difficult to track. The exception, which almost everyone has encountered, is division by zero. If it occurs during testing, it can usually be fixed with relative ease, but if it occurs when the system is in use, it can be nearly impossible to fix. Avoiding this problem is largely a matter of design and implementation. It usually creeps into the code in one of three ways:

- The equation containing the division does not have an intrinsic problem, but the way it is broken down in the code introduces the problem. In other words, to make the code simpler, some intermediate results were calculated, one of which became zero, while the full expression did not.

- The problem is intrinsic in the formula but not in the values. For example, consider sin (x)/x. For small values of x, this approaches unity, but if you attempt to compute it at zero using the formula, a division by zero error will occur. The solution is to understand how close you can get to zero without generating the fault (or losing accuracy) protecting the formula with an "*if*" statement, using a numerical approximation near zero.

- The problem is intrinsic in the formula and to the values. In other words, the problem is real. There are two possibilities. Either the formula is wrong, or the component is using it outside of the domain for which it was intended. In other words, the error condition should not occur. This later condition may indicate a problem with the requirements, or with another part of the code, that is generating the improper condition.

Other, more subtle problems with numerical accuracy include:

- ***Catastrophic subtraction*** — This occurs when two floating-point values, that are very nearly equal, are subtracted. For example, if the two values only differ in the last two bits of their mantissa, the difference will only have two bits of precision, no matter how many digits the floating-point numbers have. In other words, if the floating-point numbers are 64-bits long, only the last two will count. Rearranging the expressions so subexpressions with similar values are not subtracted is usually the solution. Of course, the formula need not have an explicit subtraction: adding a positive and a negative number can cause the problem.

- *Rounding errors* — When floating-point numbers are computed, the intermediate results computed by the run-time library will always have more precision than the data types. For example, when two 24-bit mantissa floating point numbers are multiplied, the intermediate result computed by the run-time system could have 48 bits but only 24 of them will be stored if the result is stored into another floating point number of the same size. A single instance of rounding is usually not a problem. However, if the calculation is performed repeatedly in an iterative algorithm, the cumulative effect may be substantial. Some algorithms are intrinsically numerically stable while others are not.

All of the above classes of problems are reduced in their frequency (but not eliminated) by the use of computing systems (hardware or software) that implement one of the IEEE floating point standards [IEEE, 1985], [IEEE, 1987]. The standards specify, among other things, that intermediate values be stored in 80 bits for 64-bit arguments. They also specify optimal rounding algorithms. These standards are implemented in many processors and runtime systems.

Finally, consider obtaining a package of mathematical functions with known properties [Kulisch et al., 1995].

REFERENCES

- **[Abelson, et al., 1985]** Harold Abelson, Gerald Jay Sussman, with Julie Sussman, *Structure and Interpretation of Computer Programs* (The MIT Electrical Engineering and Computer Science Series), The MIT Press, Cambridge, MA, 1985.

- **[Abrash, 1994]** M. Abrash, *Zen of Code Optimization,* Coriolis Group Books, Scottsdale, AZ, 1994.

- **[Beizer, 1983]** B. Beizer, *Software Testing Techniques,* Van Nostrand Reinhold, NY, 1983.

- **[Bentley, 1986]** J. Bentley, *Programming Pearls.* Addison Wesley, Reading, MA, 1986.

- **[Bentley, 1988]** J. Bentley, *More Programming Pearls: Confessions of a Coder.* Addison Wesley, Reading, MA, 1988.

- **[Boehm, 1981]** B. Boehm, *Software Engineering Economics*. Prentice-Hall, Englewood Cliffs, NJ, 1981.

- **[Christensen and Thayer, 2002]** M. Christensen and R. Thayer, *The Project Manager's Guide to Software Engineering's Best Practices,* IEEE, CS Press, Los Alamitos, CA, 2002.

- **[COCOMO, 2000]** *COCOMO II User's Manual*, Available from the research pages of the University of Southern California Web site, 2000.

- **[Daconta, 1993]** M. Daconta, *C Pointers and Dynamic Memory Management*, QED Publishing Group, Wellesley, MA, 1993.

- **[Dijkstra, 1968]** E. Dijkstre, *"Go To Statement Considered Harmful,"* *Comm. ACM*, Vol. 11, no. 3, 1968.

- **[Dyer, 1992]** M. Dyer, *The Cleanroom Approach to Quality Software Development*, John Wiley, NY, 1992.

- **[Grimaldi, 2000]** R. Grimaldi, *Discrete and Combinatorial Mathematics,* Addison Wesley Longman, Reading, MA, 2000.

- **[Hetzel, 1988]** B. Hetzel, *The Complete Guide to Software Testing,* QED Publishing Group, Wellesley, MA, 1988.

- **[IEEE Standard 754-1985]** *IEEE Standard 754-1985: A Standard for Binary Floating-Point Arithmetic.* Institute of Electrical and Electronic Engineers, Piscataway, NJ, 1985.

- **[IEEE Standard 830-1998]** IEEE Standard 830-1998, *IEEE Recommended Practice for Software Requirements Specifications.* IEEE, Piscataway, NJ, 1998.

- **[IEEE Standard 854-1987]** *IEEE Standard 854-1987: A Standard for Radix-Independent Floating-Point Arithmetic*, IEEE, Piscataway, NJ, 1987.

- **[IEEE Standard 1471-2000]** *IEEE Standard 1471-2000: Recommended Practice for Architectural Descriptions of Software-Intensive Systems.* IEEE, Piscataway, NJ, 2000.

- **[Jain, 1991]** R. Jain, *The Art of Computer System Performance Evaluation,* John Wiley, New York, 1991.

- **[Kant, 1992]** K. Kant, *Introduction to Computer System Performance Evaluation.* McGraw-Hill, NY, 1992.

- **[Kit, 1995]** E. Kit, *Software Testing in the Real World,* Addison Wesley, Reading, MA, 1995.

- **[Knuth, 1981]** D. Knuth, *The Art of Computer Programming* (three volumes), Addison Wesley, Reading, MA, 1981.

- **[Kruse, et al., 1991]** R. Kruse, B. Leung, C. Tondo, *Data Structures and Program Design in C*, Prentice-Hall, Englewood Cliffs, NJ, 1991.

- **[Kulisch, et al., 1995]** U. Kulisch, R. Hammer, M. Hocks, and D. Ratz, *C++ Toolbox for Verified Computation*, Springer-Verlag, Berlin, 1995.

- **[Ledgard, 1987]** H. Ledgard, *Professional Software* (two volumes), Addison Wesley, Reading, MA, 1987.

- **[Lynch, 1985]** T. Lynch, *Data Compression Techniques, and Applications*, Van Nostrand Reinhold, NY, 1985.

- **[McConnell, 1993]** S. McConnell, *Code Complete,* Microsoft Press, Redmond, WA, 1993.

- **[Myers, 1975]** G. Myers, *Reliable Software Through Composite Design.* Van Nostrand Reinhold, NY, 1975.

- **[Nemzow, 1994]** M. Nemzow, *Computer Performance Optimization,* McGraw-Hill, NY, 1994.

- **[Nemzow, 1994a]** M. Nemzow, *Enterprise Network Performance Optimization,* McGraw-Hill, NY, 1994.

- **[Oppenheim and Schafer, 1989]** A. Oppenheim and R. Schafer, *Discrete-Time Signal Processing*, Prentice-Hall, Englewood Cliffs, NJ, 1989.

- **[Phadke, 1989]** Madhav S. Phadke, *Quality Engineering Using Robust Design.* 1st ed., Prentice Hall, Upper Saddle River, NJ, 1989.

- **[Paramount Pictures, 1951]** G. Pal (producer), *When Worlds Collide*, Paramount Pictures, Hollywood, CA, 1951.

- **[Press, et al., 1990]** W. Press, B. Flannery, A. Teukosky, and W. Vetterling, *Numerical Recipes in C.* Cambridge University Press, Cambridge, UK, 1990.

- **[Pyster, 1980]** A. Pyster, *Compiler Design and Construction*, Van Nostrand Reinhold, New York, 1980.

- **[Wirth, 1976]** Niklaus Wirth, *Algorithms Data Structures = Programs*, Prentice-Hall Series in Automatic Computation, Upper Saddle River, NJ, 1976.

- **[Wirth, 1979]** Niklaus Wirth, Algorithms + Data Structures = Programs, Prentice Hall PTR Upper Saddle River, NJ, 1079

Chapter 4: Coding Standards

Presentation by Jordan Belone[4]

Coding standards, sometimes referred to as programming styles or coding conventions, are a very important asset to programmers. Unfortunately, they are often overlooked by junior as well as some experienced program developers because many of the recommended coding standards do not actually affect the compilation of the code itself, a concept that we will focus on later. This report shall try to demonstrate the importance of coding standards, and why they have already become a common practice for businesses and programming language developers alike.

I shall begin with a definition of coding standards. In general they are a set of standards and guidelines which are/should be used when writing the source code for a program. Standards are rules, which a developer is expected to follow. Guidelines however, are much less followed. They are stylistic measures, which have no direct effect on the compilation of the guide, and exist only to make the source code more humanly readable. They are optional, but highly recommended.

An important thing to remember about coding standards that many programmers seem to forget, is that coding standards are not an abstract idea. In fact, for many companies and practically all programming languages, a coding standard is a physical document, and is vital for producing uniform code.

Coding standards vary in importance between companies and programming languages. For the COBOL programmer for example, commenting code (an area of coding standards, which will be explained later) would be significantly less important than in other languages, as it is possible to make COBOL self-documenting. It also may not be important for a company that is not going to be continually producing software, and it may be more time consuming than just building the software without coding standards.

However, in open source software, the importance of coding standards cannot be emphasised enough. The most prominent example of this is Mozilla Firefox, the open source internet browser which makes all its code freely available online. This approach is taken so other developers can look over the source

---

4. I found this article on *coding standards* on my web browser. I tried to find the author to ask his permission to use his article in my small book titled *Software Construction*. I was never able to find him. All I know about the author is that he spells like a Brit. RHT

code, possibly identifying possible security flaws/bugs in the code as well as extending the code. It would be extremely difficult for other developers to identify potential flaws in the program if it did not follow a particular style convention. Also, Firefox has a huge database of extensions, some of which it created itself but many of which were created by external developers. External extensions would be very difficult to understand and integrate into Firefox if they did not follow a coding standard, and without them, Firefox may not have become such a popular internet browser as it is today.

Coding standards, when implemented correctly, can provide huge advantages to the produced code and the companies implementing them. They can make a great difference when maintaining software. Maintenance is arguably the most important step in the Software Development Life Cycle (SDLC) and is definitely the most expensive.

When the developers who are responsible for maintaining the code are different from the ones who produced it, they must understand code developed by another developer. With the lack of a coding standard, this task would be extremely different because the majority of programmers have their own programming style, which is often difficult to follow by other coders and developers. Having a coding standard, however, means that all software programmers should be able to easily read and understand the entire code written by others for a program, as long as it conforms to the coding standard and the developer who is maintaining the code is experienced with the standard.

Software that complies with a pre-existing coding standard has also been proven to contain fewer bugs and generally be of a higher quality. This is because the software is much easier to test, and any potential errors or problems with the code can be fixed in the development process of the program.

Since the cost of fixing a bug in a program increases exponentially over time, it is important to catch such bugs in the early development process to save money. This is, however, not the only way a coding standard can save money for a company. Having a pre-existing coding standard that the developers are already familiar with can speed up the overall software development process, and get the product to market faster, possibly making more money for the company. Also, as discussed earlier, software developed to a coding standard contains fewer bugs, which means less time and money has to be spent maintaining the code.

Complying with a coding standard can also have a huge effect on teamwork. Often, programmers learn to program in different ways and have varying, often conflicting, programming styles. This can cause conflict within a development team and such conflicts would certainly make the team less efficient. Having a pre-existing coding standard, however, means that there are no discrepancies as to how the program should be designed.

Another way in which teamwork is affected is the ability to team switch, i.e., programmers working on other projects transfer into other development groups. Having a company-encompassing coding standard means that developers can be switched from different parts of the company and require very little training to begin programming in other teams. This complements critical path analysis, a business technique which identifies the areas of a project which, if delayed, would delay the entire project. If such areas are identified, then developers can be easily assigned to such areas, ensuring the project is finished on time.

There are two main types of coding standards: (1) coding standards written for a company, and (2) coding standards written for a programming language. The company-coding standard is what developers are expected to follow within the company, whereas the coding standard for a programming language is what the programming language developers recommend all programmers should follow.

Often, both contain many similar points, and for companies that only program in one language, both are often almost identical. However, companies that develop software in a variety of languages often formulate their own coding standard. For example, Mozilla Firefox is written in C++, XUL, XBL, and JavaScript. Instead of inefficiently using the four separate coding standards of each language, Mozilla has created its own coding standard.

1. Examples of Coding Standards

In each coding standard, there are many points and rules. For example, the coding standard for GNU's Not UNIX (GNU) (an open source operating system) is 88 pages long. There are many points contained within such coding standards, some of which I shall discuss here.

1.1 Indentation.

Indentation is discussed in all coding standards at some level. In some languages, the compiler uses indentions to identify the scope of functions (Type 1).

In freeform languages, such as Java, indentation has no actual effect on the compilation of code itself, and is just used to make the code more human readable, giving an indication of scope without actually affecting the program directly (Type 2).

It is important to understand that Type 1 and Type 2 do exactly the same thing, and since both pieces of code are written in Java, the compiler ignores the indentation.

The indentation is meant for the human user. It can be seen when examining both pieces of code. Type 2 code is much easier to read than Type 1 code. I have used a simple *if-else loop* to demonstrate how even the simplest of code can become confusing when indentation is not used effectively.

The Type 2 code will not only be easier to read and understand for the person who is developing the code, but will make it much easier for the developer maintaining the code to do their job correctly and effectively.

1.2 Commenting code.

Commenting code is also an area, which is addressed in the majority of coding standards. Other languages, such as Tex, are self-documenting and are able to produce automatically documentation describing them. In others, such as COBOL, because the code itself is self-commenting, there is less of an emphasis on such areas. However, just commenting code is not enough, as some comments can actually be of little use at all, and can often make the code more confusing.

Comments should clearly demonstrate the function of the code, not only to other developers but also to you. Often when writing larger programs, it can be easy to lose track of what certain functions do, and it is easier to read a well written comment that it is to trawl through lines of code trying to remember the function of the code already written.

Comments should also not be too long. If code has followed a coding standard, it should not be too difficult for developers to understand the code. Long comments are often unnecessary and make the code look messy.

Every line should not necessarily contain comments.

Often lines such as *loops* and *variable* assignments are very simple to understand and do not require comments. Most coding standards recommend commenting the end of functions and objects, rather than commenting every line. In functional languages, it is generally recommended that a comment explaining a function is written before the code itself.

The comments should also not be too complicated. The point of a comment is to help the reader understand the code. It is important to keep the comments short and simple. For programs used on a global scale, it is recommended that they be written in English, as this is the most commonly used language among developers today.

According to the Ada coding standard [2005],

> *"Programmers should include comments whenever it is difficult to understand the code without the comments."*

This is very good advice, and the huge majority of other coding standards agree with this statement.

1.3 Variable declarations.

Variable declarations are another area, which is almost always included in coding standards. Variable declarations are often overlooked when programming, but in larger programs, they can result in the difference between understanding code and not understanding it.

Variable declarations should be long and demonstrate the function of the variable which they store. In the majority of programming languages, a variable can have any name, excluding a few keywords that are reserved by the language it-self.

Code that has been given generic names do not help explain the variables at all. I have used a basic "if" *loop* to demonstrate that even the most basic functions can be confusing without correct variable declarations. It must be remembered, however, that almost all programs will be more complex than this, therefore, making the code even less difficult to read. Having appropriate naming variables can reduce the need for comments in our code and make it more self-explanatory.

These are only a few of the many details generally included in coding standards. Others range from the more general "avoid big functions" to the more specific "limit the line length to a maximum of 120 characters."

2. Conclusion

It is vital to remember that coding standards are only as useful as the method which is used to enforce them. Even if a company has a well thought out, comprehensive coding standard, if it is not enforced, the code is useless. Programs such as Style Cop can be used to check code against the coding standard of the Microsoft development centre. Other tools, such as the Energy Code Analyser, can not only check that code complies with the Java coding standard, but can also detect bugs in the system. This is, however, not a fully comprehensive solution and it is ultimately the responsibility of the programmer to comply with the standard.

REFERENCES

- **[Ada coding standard 2005]** Ada 2005 is the informal name for the recent update to the Ada language. Formally, it is embodied in an Amendment to the Ada standard (ISO/IEC 8652:1995).

- **[Brian W. Kernighan and P.J. Plauger 1978]** *The Elements of Programming Style, 2nd ed.*, McGraw Hill, New York, 1978.

- **[Enerjy Code Analyser 2016]** http://www.enerjy.com/index.htm.

- **[GNU Coding Standard 2016]** http://www.gnu.org/prep/standards /standards.html.

- **[Mozilla Firefox Coding Standards 2016]** https://developer.mozilla .org /En/Developer_Guide/Coding_Style.

- **[Rob Pike and Brian W. Kernighan 1999]** *The Practice of Programming,* Addison-Wesley Professional Computing Series, 1st ed., Boston, 1999.

Chapter 5: Software Construction Exercises

These exercises are provided to encourage the student to browse the chapter looking for answers to the questions provided. If truth be told, the correct answer for all software engineering questions is "it depends." To avoid this issue, a set of possible answers is provided. There is (supposedly) only one correct answer.

If you are using this book as a textbook in a university course, your instructor may require you to justify your answer. (Why are some of the possible answers correct and why are some of them wrong?) The instructor might also ask you to identify any assumptions you depended on when arriving at your answer.

However, if you are very clever, maybe you can come up with more than one correct answer (which of course you have to justify).

1.   **The following are possible aims of the code construction phase:**
   I.   **Minimize complexity**
   II. **Minimize code size**
   III. **Anticipate change**
   IV. **Construct for verification**

   Which set of the above aims is fundamental to the software engineering construction phase?

   [a] I, II, & III
   [b] I, II, & IV
   [c] I, III, & IV
   [d] II, III, & IV

2.   **Which of the following is NOT a role of software engineering construction?**

   [a] Producing a description of software's internal construction
   [b] Verifying that the architectural design is acceptable
   [c] Designing and writing routines and modules
   [d] Tuning the code to make it shorter and/or faster

3.   **The main advantage of structured programming is:**

   [a] It is more efficient
   [b] It tends to be more reliable
   [c] It is easier to write
   [d] It can be flowcharted

4. **Construction quality assurance implies which of the following?**

   [a] The construction process is of good quality
   [b] Quality assurance techniques ensure use of coding standards
   [c] Testing is performed according to a predefined test plan
   [d] Appropriate metrics are evaluated to ensure the presence of desired quality factors

5. **Valid approaches to minimizing the complexity of code include which of the following?**

   I.   Writing readable code
   II. Writing brief code
   III. Writing efficient code
   IV. Using appropriate standards

   [a] I & II
   [b] III & IV
   [c] I & IV
   [d] II & III

6. **Which of the following represent approaches to producing high-quality code?**

   I.   Unit testing
   II. Technical reviews
   III. Use of assertions
   IV. Static analysis

   [a] I only
   [b] I & II
   [c] I, II, & III
   [d] I, II, III, & IV

7. **Software construction normally involves which two of the following types of testing?**

   I.   Integration testing
   II. Unit testing
   III. Random testing
   IV. System testing

   [a] I & II
   [b] III & IV
   [c] I & IV
   [d] II & III

8. At times there is no alternative to using a specific programming language. Which of the following are correct?:

   I. The acquisition agency specifies the language.
   II. There might be no language choice available on the target computer.
   III. The development contractor wants to learn a new language (and expects the acquisition organization to pay for it).
   IV. The acquisition management is unwilling to pay for a new compiler.
   V. An IEEE SED standard requires a specific programming standard.
   VI. The programming staff has only been trained in a specific language.

   [a] I, V, & VI
   [b] I, II, III, IV, & VI
   [c] II, IV, & V
   [d] V

9. The following are aspects of source code analysis:

   I. Program proving
   II. Anomaly analysis
   III. Statement execution frequency
   IV. Symbolic execution

   Which of the above terms are aspects of static analysis?

   [a] I, II, & III
   [b] I, III, & IV
   [c] II, III, & IV
   [d] I, II, & IV

10. Which two of the following statements are true?

   I. Coding includes unit testing.
   II. Programming includes unit testing.
   III. Coding is the process of expressing a computer program in a programming language.
   IV. Programming is the process of expressing a computer program in a programming language.

   [a] I
   [b] I, II, and IV
   [c] II & III
   [d] II & IV

11. In object-orientation, the difference between an instance and a class is:

[a] An instance declares data items that are called "attributes."
[b] An instance declares subroutines that are called "methods."
[c] An instance has memory space allocated to it; a class does not.
[d] An instance has a name that can be referred to during the design.

12. Comments in code should be designed to be which of the following?

I. A description of the code's intent
II. To make the code maintainable
III. As brief as possible

[a] I only
[b] I & II
[c] II & III
[d] I, II, & III

13. Which of the following statements is true about unit testing of object-oriented systems?

[a] Unit testing best takes place at the method level.
[b] Unit testing best takes place at the class level.
[c] Unit testing is infeasible.
[d] Unit testing focuses on testing individual attributes rather than methods.

14. If software is not modular, it will be difficult to:

I. Correct errors
II. Develop a system prototype
III. Implement reuse
IV. Make task assignments to individual software engineers

[a] I only
[b] I, III, & IV
[c] I, II, III, & IV
[d] IV only

15. Which of the following represent approaches to producing high quality code?

I. Unit testing
II. Technical reviews
III. Use of assertions
IV. Static analysis

[a] I only
[b] I and II
[c] II and III
[d] I, II, Ill and IV

16. **Outputs from the code and unit-testing phases normally include which of the following?**

[a] Code listings and test reports
[b] Code listings and a draft maintenance manual
[c] Code listings and a draft users' manual
[d] Test reports and a draft maintenance manual

17. **All of the following are major aspects of a software design EXCEPT:**

[a] Control structures
[b] Algorithms
[c] Requirements
[d] Requirements traceability matrix

18. **Attributes of software:**

I. Software has an engineering cost but not a production cost
II. Software is always assembled from existing components like hardware
III. Software is cheaper to modify than hardware
IV. Software is developed or engineered; it is not manufactured in the classical since
V. Software maintenance does NOT return the software system to its original configuration

[a] I, II, and V
[b] III and V
[c] I, IV, and V
[d] III and V

19. **Basic characteristics of good metrics include which of the following?**

I. They are meaningful, objective, and a by-product of the process
II. They display trends, identify poor processes, and are quantifiable
III. They display trends and are unambiguous
IV. They are objective, quantifiable, and consistent across all phases

[a] I and IV
[b] I and III
[c] II and III
[d] II and IV

20. Inspections have the following attributes:

I. Are led by a moderator
II. Are attended by the author's managers (different than walkthroughs)
III. A recorder records the errors found
IV. The inspection team attempts to find a solution to the errors found
V. Inspections were developed at IBM
VI. Each inspection takes a minimum of 5 hours

[a] I, III, IV, and VI
[b] I, II, III, IV, and V
[c] II, III, and VI
[d] I, III, and V

INDEX

Bjorke, Per, 9
bottom up, 92
bottom-up integration, 22, 23, 24
Buckley Fletcher, 9

catastrophic subtraction, 98
Christensen, Mark, v, x, 47
code complete, x, 3, 36, 39, 41, 42, 44, 45, 46, 100
code development, 31
code stepping, 21
code tuning, 17, 32
coding standard, 102, 103, 104, 105, 106
coding, v, 4, 16, 17, 18, 44, 46, 50, 102, 103, 104, 106, 110
cohesion, 65, 83
commenting code, 105
communication methods, 6
complexity, minimizing, 4, 5
components, 30, 79, 91
computing platform, 32
conciseness, 54, 55
concurrency primitives, 4, 29
configuration, run-time, 4, 28
consistency, 53, 54, 55
constructing for verification, 4, 6
construction, i, ii, iii, v, x, 1, 2, 3, 4, 5, 8, 9, 13, 14, 15, 18, 19, 20, 24, 30, 33, 39, 41, 42, 44, 45, 46, 47, 49, 67, 72, 73, 101, 102, 107, 108
construction planning, major issues in, 4, 14
construction quality, improving, 22, 23
construction, software, 1, 2, 3, 40, 45, 108construction techniques, table-driven, 26
construction technologies, 2, 4, 24
containers, 30
co-specification, 31
coupling, 5, 7, 60, 62, 65, 66, 67, 76, 77, 78, 82, 84

data abstraction, 25
data design, 15
defensive programming, 4, 27, 77
design by contract, 27
design evaluation, v, 56
design, software, 15
design, the internals, 78
desktop applications, 49

Professional Software Engineering Master Certification, 1
profiling, software, 35
program slicing, 35
programming, vi, 1, 2, 7, 8, 11, 16, 24, 25, 26, 27, 28, 29, 32, 33, 34, 39, 44, 45, 46, 50, 51, 52, 67, 69, 78, 79, 82, 83, 86, 102, 103, 104, 105, 106, 107, 110
programming languages, 6, 16project management plans, 19

quality, software, 2

refactoring, 33, 34, 45
reuse, external, 7
reuse, internal, 7
reuse, software engineering, 6
rounding errors, 98
Royce, W.W., 22, 37
run-time configuration, 4, 28

schedule network, 73
semaphore, 29
slicing, program, 35
software construction, 1, 2, 3, 40, 45, 108
software construction fundamentals, 2
software construction tools, 2
software design, 15
software engineering construction (SEC), 1, 39
software library, 7
software profiling, 35
software quality, 20
staged-delivery model, 11
standards in construction, 4, 8
standards, external, 8
standards, internal, 9
suitability, 59, 62
systems and software engineering, 8

table-driven construction techniques, 26
test-first programming, 4, 33
testing, pros and cons of integration, 94
testing, unit, 4, 18, 34, 40, 46, 108, 111
testing, pair-wise, 89
testing, white box, 34thread of execution, 92
Tockey, Steve, xii

tool kit languages, 16
tools, v, 4, 6, 24, 33, 45
top down, 91
top-down integration, 23
tuning, numeric accuracy, 98
Twelve Principles of Agile Processes, 12

unit test cases, 87
unit testing, 4, 18, 34, 40, 46, 108, 111

visual notations, 16

walkthrough, 61
widget (a.k.a. control), 34

NOTES

NOTES

NOTES

NOTES

NOTES

NOTES

NOTES

www.ingramcontent.com/pod-product-compliance
Lightning Source LLC
Chambersburg PA
CBHW080559220326
41599CB00032B/6539